# The IMA Volumes in Mathematics and its Applications

## Volume 60

*Series Editors*
Avner Friedman Willard Miller, Jr.

# Institute for Mathematics and its Applications
## IMA

The **Institute for Mathematics and its Applications** was established by a grant from the National Science Foundation to the University of Minnesota in 1982. The IMA seeks to encourage the development and study of fresh mathematical concepts and questions of concern to the other sciences by bringing together mathematicians and scientists from diverse fields in an atmosphere that will stimulate discussion and collaboration.

The IMA Volumes are intended to involve the broader scientific community in this process.

Avner Friedman, Director
Willard Miller, Jr., Associate Director

* * * * * * * * * *

## IMA ANNUAL PROGRAMS

1982–1983 **Statistical and Continuum Approaches to Phase Transition**
1983–1984 **Mathematical Models for the Economics of Decentralized Resource Allocation**
1984–1985 **Continuum Physics and Partial Differential Equations**
1985–1986 **Stochastic Differential Equations and Their Applications**
1986–1987 **Scientific Computation**
1987–1988 **Applied Combinatorics**
1988–1989 **Nonlinear Waves**
1989–1990 **Dynamical Systems and Their Applications**
1990–1991 **Phase Transitions and Free Boundaries**
1991–1992 **Applied Linear Algebra**
1992–1993 **Control Theory and its Applications**
1993–1994 **Emerging Applications of Probability**

## IMA SUMMER PROGRAMS

1987 **Robotics**
1988 **Signal Processing**
1989 **Robustness, Diagnostics, Computing and Graphics in Statistics**
1990 **Radar and Sonar**
1990 **Time Series**
1991 **Semiconductors**
1992 **Environmental Studies: Mathematical, Computational, and Statistical Analysis**

* * * * * * * * * *

## SPRINGER LECTURE NOTES FROM THE IMA:

*The Mathematics and Physics of Disordered Media*
Editors: Barry Hughes and Barry Ninham
(Lecture Notes in Math., Volume 1035, 1983)

*Orienting Polymers*
Editor: J.L. Ericksen
(Lecture Notes in Math., Volume 1063, 1984)

*New Perspectives in Thermodynamics*
Editor: James Serrin
(Springer-Verlag, 1986)

*Models of Economic Dynamics*
Editor: Hugo Sonnenschein
(Lecture Notes in Econ., Volume 264, 1986)

Gene Golub Anne Greenbaum
Mitchell Luskin
Editors

# Recent Advances in Iterative Methods

With 45 Illustrations

Springer-Verlag
New York Berlin Heidelberg London Paris
Tokyo Hong Kong Barcelona Budapest

Gene Golub
Department of Computer Science
Stanford University
Stanford, CA 94035 USA

Anne Greenbaum
Courant Institute of Mathematical Sciences
251 Mercer Street
New York, NY 10012 USA

Mitchell Luskin
School of Mathematics
University of Minnesota
127 Vincent Hall, 206 Church Street SE
Minneapolis, MN 55455 USA

*Series Editors:*
Avner Friedman
Willard Miller, Jr.
Institute for Mathematics and its Applications
University of Minnesota
Minneapolis, MN 55455 USA

Mathematics Subject Classifications (1991): 65-03, 65-06, 65F10, 65F15, 65F25, 65F30, 65F50, 65Y05, 65Y10, 65Y20

Library of Congress Cataloging-in-Publication Data
Recent advances in iterative methods / Gene Golub, Anne Greenbaum, Mitchell Luskin [editors]
p. cm. – (The IMA volumes in mathematics and its applications)
Papers from the IMA Workshop on Iterative Methods for Sparse and Structured Problems, held in Minneapolis, Minn., Feb. 24–Mar. 1, 1992.
Includes bibliographical references.
ISBN 0-387-94252-1 (New York : acid-free). – ISBN 3-540-94252-1 (Berlin : acid-free)
1. Iterative methods (Mathematics)–Congresses. I. Golub, Gene H. (Gene Howard), 1932–. II. Greenbaum, Anne. III. Luskin, Mitchell Barry, 1951–. IV. IMA Workshop on Iterative Methods for Sparse and Structured Problems (1992:Minneapolis, Minn.). V. Series.
QA297.8.R43 1994 93-50626
511′.4–dc20

Printed on acid-free paper.


Production managed by Hal Henglein; manufacturing supervised by Jacqui Ashri.
Camera-ready copy prepared by the IMA.
Printed and bound by Edwards Brothers, Inc., Ann Arbor, MI.
Printed in the United States of America.

9 8 7 6 5 4 3 2 1

ISBN 0-387-94252-1 Springer-Verlag New York Berlin Heidelberg
ISBN 3-540-94252-1 Springer-Verlag Berlin Heidelberg New York

# The IMA Volumes
## in Mathematics and its Applications

### Current Volumes:

Volume 1: Homogenization and Effective Moduli of Materials and Media
Editors: Jerry Ericksen, David Kinderlehrer, Robert Kohn, and J.-L. Lions

Volume 2: Oscillation Theory, Computation, and Methods of Compensated Compactness
Editors: Constantine Dafermos, Jerry Ericksen, David Kinderlehrer, and Marshall Slemrod

Volume 3: Metastability and Incompletely Posed Problems
Editors: Stuart Antman, Jerry Ericksen, David Kinderlehrer, and Ingo Muller

Volume 4: Dynamical Problems in Continuum Physics
Editors: Jerry Bona, Constantine Dafermos, Jerry Ericksen, and David Kinderlehrer

Volume 5: Theory and Applications of Liquid Crystals
Editors: Jerry Ericksen and David Kinderlehrer

Volume 6: Amorphous Polymers and Non-Newtonian Fluids
Editors: Constantine Dafermos, Jerry Ericksen, and David Kinderlehrer

Volume 7: Random Media
Editor: George Papanicolaou

Volume 8: Percolation Theory and Ergodic Theory of Infinite Particle Systems
Editor: Harry Kesten

Volume 9: Hydrodynamic Behavior and Interacting Particle Systems
Editor: George Papanicolaou

Volume 10: Stochastic Differential Systems, Stochastic Control Theory and Applications
Editors: Wendell Fleming and Pierre-Louis Lions

Volume 11: Numerical Simulation in Oil Recovery
Editor: Mary Fanett Wheeler

Volume 12: Computational Fluid Dynamics and Reacting Gas Flows
Editors: Bjorn Engquist, M. Luskin, Andrew Majda

Volume 13: Numerical Algorithms for Parallel Computer Architectures
Editor: Martin H. Schultz

Volume 14: Mathematical Aspects of Scientific Software
Editor: J.R. Rice

Volume 15: Mathematical Frontiers in Computational Chemical Physics
Editor: D. Truhlar

Volume 16: Mathematics in Industrial Problems
by Avner Friedman

Volume 17: Applications of Combinatorics and Graph Theory to the Biological and Social Sciences
Editor: Fred Roberts

Volume 18: $q$-Series and Partitions
Editor: Dennis Stanton

Volume 19: Invariant Theory and Tableaux
Editor: Dennis Stanton

Volume 20: Coding Theory and Design Theory Part I: Coding Theory
Editor: Dijen Ray-Chaudhuri

Volume 21: Coding Theory and Design Theory Part II: Design Theory
Editor: Dijen Ray-Chaudhuri

Volume 22: Signal Processing: Part I - Signal Processing Theory
Editors: L. Auslander, F.A. Grünbaum, J.W. Helton, T. Kailath, P. Khargonekar, and S. Mitter

Volume 23: Signal Processing: Part II - Control Theory and Applications of Signal Processing
Editors: L. Auslander, F.A. Grünbaum, J.W. Helton, T. Kailath, P. Khargonekar, and S. Mitter

Volume 24: Mathematics in Industrial Problems, Part 2
by Avner Friedman

Volume 25: Solitons in Physics, Mathematics, and Nonlinear Optics
Editors: Peter J. Olver and David H. Sattinger

Volume 26: Two Phase Flows and Waves
Editors: Daniel D. Joseph and David G. Schaeffer

Volume 27: Nonlinear Evolution Equations that Change Type
Editors: Barbara Lee Keyfitz and Michael Shearer

Volume 28: Computer Aided Proofs in Analysis
Editors: Kenneth Meyer and Dieter Schmidt

Volume 29: Multidimensional Hyperbolic Problems and Computations
Editors: Andrew Majda and Jim Glimm

Volume 30: Microlocal Analysis and Nonlinear Waves
Editors: Michael Beals, R. Melrose, and J. Rauch

Volume 31: Mathematics in Industrial Problems, Part 3
by Avner Friedman

Volume 32: Radar and Sonar, Part I
by Richard Blahut, Willard Miller, Jr., and Calvin Wilcox

Volume 33: Directions in Robust Statistics and Diagnostics: Part I
Editors: Werner A. Stahel and Sanford Weisberg

Volume 34: Directions in Robust Statistics and Diagnostics: Part II
Editors: Werner A. Stahel and Sanford Weisberg

Volume 35: Dynamical Issues in Combustion Theory
Editors: P. Fife, A. Linan, and F.A. Williams

Volume 36: Computing and Graphics in Statistics
Editors: Andreas Buja and Paul Tukey

Volume 37: Patterns and Dynamics in Reactive Media
Editors: Harry Swinney, Gus Aris, and Don Aronson

Volume 38: Mathematics in Industrial Problems, Part 4
by Avner Friedman

Volume 39: Radar and Sonar, Part II
Editors: F. Alberto Grünbaum, Marvin Bernfeld, and
Richard E. Blahut

Volume 53: Variational and Free Boundary Problems
Editors: Avner Friedman and Joel Spruck

Volume 54: Microstructure and Phase Transition
Editors: David Kinderlehrer, Richard James, Mitchell Luskin, and Jerry L. Ericksen

Volume 55: Turbulence in Fluid Flows: A Dynamical Systems Approach
Editors: George R. Sell, Ciprian Foias, and Roger Temam

Volume 56: Graph Theory and Sparse Matrix Computation
Editors: Alan George, John R. Gilbert, and Joseph W.H. Liu

Volume 57: Mathematics in Industrial Problems, Part 6
by Avner Friedman

Volume 58: Semiconductors, Part I
Editors: W.M. Coughran, Jr., Julian Cole, Peter Lloyd, and Jacob White

Volume 59: Semiconductors, Part II
Editors: W.M. Coughran, Jr., Julian Cole, Peter Lloyd, and Jacob White

Volume 60: Recent Advances in Iterative Methods
Editors: Gene Golub, Anne Greenbaum, and Mitchell Luskin

**Forthcoming Volumes**:

Hamiltonian Dynamical Systems

1991-1992: *Applied Linear Algebra*

Linear Algebra for Signal Processing

1992 Summer Program: *Environmental Studies*

Environmental Studies

1992–1993: *Control Theory*

Robust Control Theory

Control and Optimal Design of Distributed Parameter Systems

Flow Control

Robotics

Nonsmooth Analysis and Geometric Methods in Deterministic Optimal Control

Systems and Control Theory for Power Systems

Adaptive Control, Filtering and Signal Processing

Discrete Event Systems, Manufacturing, Systems, and Communication Networks

Mathematical Finance

Summer Program: *Modeling, Mesh Generation, and Adaptive Numerical Methods for Partial Differential Equations*

1993-1994: *Emerging Applications of Probability*

Probability and Algorithms

Finite Markov Chain Renaissance

Random Discrete Structures

## FOREWORD

This IMA Volume in Mathematics and its Applications

## RECENT ADVANCES IN ITERATIVE METHODS

is based on the proceedings of a workshop that was an integral part of the 1991-92 IMA program on "Applied Linear Algebra." Large systems of matrix equations arise frequently in applications and they have the property that they are sparse and/or structured. The purpose of this workshop was to bring together researchers in numerical analysis and various application areas to discuss where such problems arise and possible methods of solution. The last two days of the meeting were a celebration dedicated to Gene Golub on the occasion of his sixtieth birthday, with the program arranged by Jack Dongarra and Paul van Dooren. We are grateful to Richard Brualdi, George Cybenko, Alan George, Gene Golub, Mitchell Luskin, and Paul Van Dooren for planning and implementing the year-long program. We especially thank Gene Golub, Anne Greenbaum, and Mitchell Luskin for organizing this workshop and editing the proceedings.

The financial support of the National Science Foundation and the Minnesota Supercomputer Institute made the workshop possible.

Avner Friedman

Willard Miller, Jr.

# PREFACE

The solution of very large linear algebra problems is an integral part of many scientific computations. For example, the numerical solution of partial differential equations by finite difference or finite element methods often requires solving systems of linear equations or eigenvalue problems in which the coefficient matrix is large but sparse (contains many zero elements). Integral equations modeling various physical phenomena are solved similarly by replacing the integrals by a quadrature formula, resulting in a large system of linear algebraic equations. In this case the matrices that arise are no longer sparse, but they are highly structured, with all $n^2$ entries of an $n$ by $n$ matrix typically being functions of just $n$ numbers. Other examples of such structured matrices include Toeplitz and Hankel matrices, which arise, for example, in signal processing applications. Similarly many statistical calculations require the solution of large linear systems or least squares problems in which the coefficient matrix has very special properties.

For many of these applications it is not feasible, in terms of computation time or memory, to store the entire matrix and solve the linear algebra problem directly. Instead, iterative methods are used. These methods do not require storage of the entire matrix, but rely instead on the ability to compute matrix vector products. This can be accomplished by storing a much smaller set of numbers (e.g., the nonzero entries of a sparse finite element or finite difference matrix) or, in cases in which there is a simple formula for the matrix-vector product, perhaps no matrix entries at all.

In recent years much research has focussed on the efficient solution of such linear algebra problems using iterative methods. The preconditioned conjugate gradient method, for example, has proved to be a very effective technique for solving symmetric positive definite linear systems and it is now in wide use throughout government laboratories and industry, as well as academia. Similarly, the Lanczos algorithm for computing eigenvalues and eigenvectors of symmetric matrices is implemented in many widely used commercial codes today. Current research issues involving the conjugate gradient algorithm include finding efficient preconditioners, estimating the error in the computed solution, and better understanding the effects of finite precision arithmetic. Several of these questions are addressed in this volume.

The state of the art for nonsymmetric problems is not so advanced. Numerous conjugate-gradient-like methods have been developed for nonsymmetric problems, but no one method has stood out as clearly superior to the others. The basic difficulty is that, unlike the symmetric case, a short recurrence for the optimal approximation from a Krylov space does not exist for most nonsymmetric problems. Consequently, one must either generate a sub-optimal approximate solution or use a long and costly recurrence to determine the optimal approximation. Clever strategies have been devised in recent years to generate an almost-optimal approximation using a short recurrence, and several of those strategies are discussed here. Another interesting research question associated with nonsymmetric matrix iterations is what determines their convergence rate. For symmetric matrices, or, more generally, for normal matrices, it is the eigenvalues that determine the behavior of iterative methods. This is not the case for nonnormal matrices, and it is not clear exactly what properties of a nonnormal matrix are required to ensure rapid convergence of iterative methods. Finally, while the particular choice of iteration strategy is of some importance, it is often even more important to determine a good preconditioner for the problem. Theoretical issues about the convergence rate of iterative methods as well as more practical issues of effective preconditioners are addressed in several of the papers contained herein.

The relative merits of various computational algorithms cannot be discussed without reference to both the problems they are intended to solve and the computers on which they are intended to run. As computer architectures advance, the comparison of different methods may change. The latest advance in architecture is toward parallel machines, and this introduces the question of efficient parallel iterative methods. Most iterative methods require computation of inner products, which may be expensive on distributed memory parallel machines. In response to this, algorithms are being developed which require fewer inner products. In addition, the preconditioning phase of an iterative method may be difficult to parallelize. The development of efficient parallel preconditioners is another research issue being addressed.

The IMA Workshop on Iterative Methods for Sparse and Structured Problems brought together researchers interested in all of these topics and more. The papers in this volume represent the latest results of mathematical and computational research into the development and analysis of robust iterative methods for the numerical solution of a variety of linear algebra problems.

Gene Golub
Anne Greenbaum
Mitchell Luskin

## CONTENTS

# SOME THEMES IN GENE H. GOLUB'S WORK ON ITERATIVE METHODS

DIANNE P. O'LEARY*

**Abstract.** Gene H. Golub has authored nearly seventy publications related to iterative methods for solving sparse matrix problems and sparse systems of nonlinear equations. Over the past 30 years, his work has delineated some of the major research directions in these fields, and his visibility and clear exposition of ideas have stimulated an extraordinary amount of interest and research activity by others. In this paper, we review these contributions.

**1. Introduction.** On the occasion of Gene Golub's 60th birthday, one day of the IMA Workshop on Iterative Methods was devoted to a celebration of his achievements. A complete review of his work, of course, is premature, since Gene Golub is still an active researcher, but we will discuss in this paper some themes from his work thus far on iterative methods.

No summary of Golub's contributions would be complete without at least a brief mention of his other professional activities. His *service to the profession* has included the founding of journals, service to SIAM as president and in many other capacities, creation of NA-NET (an electronic newsletter and mail forwarding service), encouragement of young colleagues, the training of 20 Ph.D. students, outstanding teaching, and textbook authorship. He has made contributions to *computational statistics* in the areas of Markov chains, maximum likelihood estimation, generalized cross validation, updating sample variance, weighted least squares, total least squares, separable nonlinear least squares, and treatment of outliers. In the field of *optimization*, his work has focused on Chebyshev overdetermined systems, linear programming, and updating matrix factorization. His work on *eigenproblems and singular value problems* has produced algorithms for the singular value decomposition, bounds for the $LR$ algorithm, understanding of the Lanczos algorithm, and algorithms for modified matrix eigenproblems.

Each of these sets of accomplishments merits its own survey paper, but here we focus on Golub's work on iterative methods.

**2. Iterating to success.** A central theme of Gene Golub's work has been the understanding of convergence properties of iterative methods and the development of algorithmic improvements for iterations.

In 1959, Gene Golub completed his Ph.D. thesis [35]. Since Richard Varga had some similar results, the work was published jointly in 1961 in a

* Computer Science Department and Institute for Advanced Computer Studies; University of Maryland; College Park, Maryland 20742. This work was completed while the author was in residence at the Institute for Mathematics and Its Applications, University of Minnesota, supported by the Office of Graduate Studies and Research, General Research Board, University of Maryland College Park, and by the IMA.

two part paper [67], discussing relations among the Richardson, SOR, and Chebyshev algorithms. The key observation was that the SOR algorithm applied to the linear system

$$\begin{bmatrix} 0 & B \\ B & 0 \end{bmatrix} \begin{bmatrix} x \\ y \end{bmatrix} + \begin{bmatrix} g \\ g \end{bmatrix} = \begin{bmatrix} x \\ y \end{bmatrix}$$

has the same form as Chebyshev or Richardson on the system $(I - B)x = g$ if $B$ is symmetric and primitive. Among these algorithms, the Chebyshev iteration operator has the smallest spectral norm. Golub and Varga derived a special form of the Chebyshev iteration for 2-cyclic matrices that requires only half the work per iteration of the standard algorithm. This idea of reducing the problem size by *cyclic reduction* was used later by Reid for the conjugate gradient iteration and by Bonhoure, Dallery, and Stewart for Markov chains.

The Golub and Kahan ('65) [45] work on algorithms for computing the singular value decomposition included the observation that the bidiagonal form for computing the SVD could be generated by applying the Lanczos algorithm to

$$\begin{bmatrix} 0 & A \\ A^* & 0 \end{bmatrix}.$$

Lanczos-based SVD algorithms have become important for sparse matrices.

Concus and Golub ('76) [15] considered linear systems with matrices of the form $A = M$ - $N$, where $M = (A + A^T)/2$ , the symmetric part of the matrix, is symmetric positive definite. They derived a conjugate-gradient-like iteration and derived some convergence results for it. This was an important step in motivating the further study of conjugate gradient algorithms for non-symmetric problems.

Golub and Overton ('81) [56] considered algorithms in the family of "inner-outer" iterations. For example, the Richardson iteration with an inexact $M$ has the form

$$x_{k+1} = x_{k-1} + \omega(\alpha z_k + x_k - x_{k-1})$$

where

$$Mz_k = (b - Ax_k) + q_k, \quad ||q_k|| \le \delta ||b - Ax_k||.$$

In the usual Richardson method, $q_k$ would be zero. They derived conditions on the size of $q_k$ to guarantee convergence of the iteration. A similar analysis including the Chebyshev iteration was described in Golub and Overton ('88) [57].

The following papers are also related to the theme of understanding and improving iterative methods:

Golub ('62) [36]: analyzed round-off error in the Richardson iteration.

Golub and Wilkinson ('65,'66) [68,69]: analyzed iterative refinement.

Anderssen and Golub ('72) [1]: studied the Richardson iteration.

Concus and Golub ('73) [14]: analyzed a preconditioned Chebyshev iteration (See §3).

Concus, Golub, and O'Leary ('76) [17]: studied the preconditioned conjugate gradient iteration.

Concus, Golub, and O'Leary ('78) [18]: applied preconditioned conjugate gradients to nonlinear elliptic equations.

Golub ('85): [41] determined error bounds for various iterative methods.

Golub and Kannan ('86) [46]: studied the nonlinear Richardson iteration.

Golub and Kent ('89) [49]: studied the eigenvalues of iteration matrices.

Gander, Golub, and Gruntz ('90) [34]: discussed extrapolation methods for solving linear systems.

Golub and O'Leary ('89) [55]: discussed the early history of the conjugate gradient and Lanczos algorithms.

Golub and De Pillis ('90) [42]: studied a two-parameter SOR algorithm.

Freund, Golub, and Nachtigal ('92) [33]: surveyed Krylov subspace methods for nonsymmetric problems.

**3. The fast (direct) track (fast and direct is not enough) .** Golub had worked on two kinds of fast direct methods for the solution of Poisson's equation over unions of rectangles: cyclic reduction was investigated by Buzbee, Golub, and Nielson [11] and capacitance matrix techniques were studied by Buzbee, Dorr, George, and Golub [10]. He became interested in using algorithms based on cyclic reduction to produce efficient solvers for more general regions and more general elliptic equations.

Concus and Golub ('73) [14] considered the Chebyshev iteration for solving general elliptic equations, preconditioning the problem by fast direct solution of the Helmholtz equation on a rectangle. They showed that the convergence rate was independent of the mesh size if the coefficients are smooth.

This idea of using fast direct methods as preconditioners recurs in later papers. Concus, Golub, and O'Leary ('76) [17] used them for conjugate gradient iterations. Golub and Mayers ('83) [52] constructed preconditioners through domain decomposition, comparing various methods. Concus, Golub, and Meurant ('85) [16] constructed block preconditioners, using block ICLU with polynomial expansion and other techniques to approximate the inverse of the tridiagonal blocks on the main diagonal.

In Elman and Golub ('89, '92) [28,25,26,27,29], cyclic reduction is revisited. They considered a linear system of the form

$$\begin{bmatrix} M_1 & N_1 \\ N_2 & M_2 \end{bmatrix} \begin{bmatrix} x_1 \\ x_2 \end{bmatrix} = \begin{bmatrix} b_1 \\ b_2 \end{bmatrix},$$

and reduced it to a problem involving the Schur complement of the matrix:

$$(M_2 - N_2 M_1^{-1} N_1) x_2 = \hat{b}_2.$$

For convection-diffusion equations, $M$ is diagonal and the reduced matrix is often symmetrizable. They reordered the unknowns, applied the block Jacobi splitting, and analyzed the convergence properties by noting that reduced problem is also a discretization of the same partial differential equation.

**4. Crucial moments.** A major theme in Golub's work on the Lanczos algorithm is the relation among Krylov sequences, orthogonal polynomials, and matrix moments. Given a symmetric matrix $A$, we can expand an arbitrary vector $r$ in terms of the eigenvectors $u_i$ of the matrix:

$$r = \sum \alpha_i u_i.$$

The *moments* of the matrix are defined as

$$\mu_m = (A^m r, r) = \int_a^b \lambda^m d\,\alpha(\lambda),$$

where $\alpha(\lambda)$ is a weight function that jumps by $\alpha_i$ at the $i$th eigenvalue $\lambda_i$ of $A$. We note that the matrix moments are defined in terms of elements in the Krylov sequence $\{r, Ar, ...A^k r\}$. The connection between moments and the Lanczos algorithm comes from the observation that if $r$ is chosen to be the initial residual of a linear system $r = b - Ax$, and we apply the Lanczos (conjugate gradient) algorithm to determine the solution, the error at the $k$th step can be defined as $||x_k - x^*||^2 = \mu_{-2}$.

Dahlquist, Eisenstat, and Golub ('72) [20] obtained bounds on the size of the error using this moment relation. Noting that

$$\mu_{-2} = \int_a^b \lambda^{-2} d\,\alpha(\lambda),$$

they found two quadrature formulas:

$$\sum \gamma_i t_i^m + \beta z$$

that reproduce $\mu_m$, $0 \le m \le 2k$, one for $z = a$ and one for $z = b$. They used these formulas to obtain upper and lower bounds on $\mu_{-2}$.

A series of papers by DeBoor and Golub ('78) [21], Golub and Kautsky ('83) [47,48], and Boley and Golub ('77, '78, '84, '87) [5,6,8,9] deals with inverse matrix eigenvalue problems. In a typical example, we are given the eigenvalues of a tridiagonal matrix $J$, the eigenvalues of its leading principal minor, and one additional piece of information, and we wish to determine the entries of $J$. Many such problems can be easily and cleverly solved using the Lanczos algorithm. Such inverse eigenvalue problems are interesting in their own right, but have application in deriving Gaussian quadrature formulas of various types.

Ferng, Golub, and Plemmons ('91) [30] discuss recursive estimation of condition numbers when a matrix is updated. They use the Lanczos algorithm to perform the update efficiently.

Elhay, Golub, and Kautsky ('91) [22,23] derived algorithms for up- and down-dating of orthogonal polynomials, in order to update solutions to least squares problems when new data points are added.

The following papers also concern Krylov sequences, orthogonal polynomials, and matrix moments:

Golub and Robertson ('67) [61]: generalized the Bairstow method to find roots of linear combinations of polynomials satisfying three-term recurrence relations.

Golub, Underwood, and Wilkinson ('72) [64]: proposed the use of the Lanczos algorithm for matrix pencils.

Golub ('73) [37]: gave a clear exposition of the Lanczos algorithm.

Golub ('74) [39]: surveyed methods for eigenvalue computations.

Golub ('74) [38]: gave bounds for matrix moments.

Golub('75) [40]: surveyed the Lanczos algorithm.

Berry and Golub ('91) [2]: considered the problem of estimating singular values using modified moments.

Boley, Elhay, Golub, and Gutknecht ('91) [4], Golub and Gutknecht ('89) [43]: discussed the computation of orthogonal polynomials with indefinite weights.

Boley, Brent, Golub, and Luk ('92) [3]: used the Lanczos algorithm in manipulating checksums to detect and correct errors in the computation of matrix operations.

Fischer and Golub ('92) [32]: discussed the computation of polynomials orthogonal over several intervals.

Elhay, Golub, and Kautsky ('92) [24]: studied tridiagonal matrices generated by sums of weight functions.

**5. Block parties.** Another significant theme in Golub's work has been the use of block algorithms, those that carry along blocks of vectors or that partition the matrix into sub-blocks. We have already discussed the Concus, Golub, and Meurant ('85) [16] work on block preconditioners, but there are other examples as well.

Golub and Underwood ('77) [63] studied the block Lanczos algorithm for finding eigenvalues of matrices. The Lanczos algorithm is performed on $p$ vectors at a time, decomposing a matrix $A$ as $AV = VT + E$, where $T$ is block tridiagonal, with $p \times p$ blocks and $E$ has rank $p$. The standard Lanczos method does not determine the multiplicity of eigenvalues, but the block algorithm can be used to determine invariant subspaces for eigenvalues up to multiplicity $p$. The blocksize $p$ can also be adjusted in order to speed convergence by exploiting gaps in the spectrum. Such block algorithms have advantages for hierarchical storage and parallel computer architectures, since one access of the matrix $A$ produces several matrix-vector products.

Cottle, Golub, and Sacher ('78) [19] considered block partitioned linear complementarity problems: find $z$ and $q$ to satisfy

$$q + Mz \geq 0 \;,\;\; z \geq 0 \;,\;\; z^T(q + Mz) = 0.$$

They assumed that $M$ was symmetric and positive definite, making the problem equivalent to a quadratic programming problem with nonnegativity constraints on the variables. They proposed a modified block SOR algorithm, in which smaller linear complementarity problems were solved in order to determine search directions. The step length was then modified if necessary in order to preserve the condition $z \geq 0$. They proved convergence of the algorithm and applied it to a journal bearing problem.

Golub, Luk, and Overton ('81) [50] developed a block Lanczos algorithm for computing singular values, applying the block Lanczos algorithm (implicitly) to

$$\begin{bmatrix} 0 & A \\ A^* & 0 \end{bmatrix}$$

Ferziger, Golub, and Thompson ('87) [31] used the block SOR algorithm on cyclically reduced convection-diffusion equations.

Golub, Plemmons, and Sameh ('88) [60] studied parallel block schemes for large scale least-squares computations.

**6. Applying himself.** Golub's work is not done in the abstact. He is most often motivated by applications problems, particularly those in engineering and statistics.

Golub and Senata ('73) [62] proved monotonic convergence of an algorithm for finding the stationary vector of an infinite Markov chain. They made use of a sequence of finite approximations from upper left hand truncations of the matrix. The standard proof of convergence relies on irreducibility of the infinite chain, but in this case they proved convergence under an alternate condition: the matrix has one column of entries bounded away from zero.

Cline, Golub, and Platzman ('76) [12] computed normal modes of oceans by computing interior eigenvalues of a matrix of order 1919 by inverse iteration, solving the linear systems by the Lanczos or `Symmlq` algorithms.

Golub, Heath, and Wahba ('79) [44] developed the generalized cross-validation approach for choosing a regularization parameter for an ill-posed problem. Their algorithm determines a parameter $\rho$ for

$$\min_x ||Ax - b||^2 + \rho||Lx||^2$$

in a way that is optimal in some sense.

Boley and Golub ('84) [7] noted that since the controllability subspace for $\dot{x} = Ax + Bu$ is the smallest invariant subspace of $A$ containing the span of the columns of $B$, the subspace can be determined iteratively by applying the block Lanczos algorithm to the matrix $A$ using the columns of $B$ as starting vectors.

Comon and Golub ('90) [13] gave a survey of some numerical techniques in signal processing. Among other ideas, they propose the Lanczos and block Lanczos algorithms for updating the signal subspace in direction-of-arrival problems.

The following papers are related to applications of iterative methods in statistics, materials science, and operations research:

Golub and Pereyra ('73, '76) [58,59]: developed algorithms for separable least squares problems.

Cottle, Golub, and Sacher ('78) [19]: proposed algorithms for complementarity problems (discussed in §5).

Pagano, Golub, and Luk ('79) [70]: proposed direct and iterative methods for solving a structured least squares photogrammetry problem.

Golub and Nash ('82) [54]: performed analysis of variance by preconditioned conjugate gradients.

Golub, Manneback, and Toint ('86) [51]: solved a geodetic least squares problem.

**7. The shoe salesman.** Golub's impact on the field would have been far less had he not put considerable effort into popularizing efficient and stable numerical algorithms. He distributed computer code for many of his algorithms, including those for separable least squares, the matrix Sylvester equation, and the block Lanczos iteration. His lecture notes [53] and textbooks, especially Golub and Van Loan ('83, '89) [65,66], have greatly increased the numerical sophistication of engineers and scientists. By founding and editing journals, he has ensured that algorithmically-oriented papers have high-quality outlets. Finally, his devotion to teaching, his lectures worldwide, his organization of cross-disciplinary meetings, and his training of graduate students have given his efforts a multiplier effect.

**8. Conclusions.** As we noted at the onset, this survey is premature (and probably already out-of-date). It is clear, though, that Gene Golub's work has had a major impact on the understanding of iterative methods, the efficient numerical solution of elliptic partial differential equations, the broadening of the applications of the Lanczos algorithm, the development of block algorithms, and the transfer of these ideas to applications.

Gene's friends will notice that the section headings of this paper could serve, after suitable permutation, as chapter headings for a biography of Gene. We will all be looking forward to upcoming chapters.

## REFERENCES


[1] R. S. Anderssen and G. H. Golub. Richardson's non-stationary matrix iterative procedure. Technical Report STAN-CS-72-304, Stanford University Computer Science Department Report, Stanford, California, 1972.

[2] M. Berry and G. H. Golub. Estimating the largest singular values of large sparse matrices via modified moments. *Numerical Algorithms*, 1:353–374, 1991.

[3] D. L. Boley, R. P. Brent, G. H. Golub, and F. T. Luk. Algorithmic fault tolerance using the Lanczos method. *SIAM J. on Matrix Anal. and Applics.*, 13:312–332, 1992.

[4] D. L. Boley, S. Elhay, G. H. Golub, and M. H. Gutknecht. Nonsymmetric Lanczos and finding orthogonal polynomials associated with indefinite weights. *Numerical Algorithms*, 1:21–43, 1991.

[5] D. L. Boley and G. H. Golub. Inverse eigenvalue problems for band matrices. In G. A. Watson, editor, *Numerical Analysis Proceedings, Biennial Conference, Dundee 1977*, pages 23–31. Springer Lecture Notes in Math. 630, New York, 1978.

[6] D. L. Boley and G. H. Golub. The matrix inverse eigenvalue problem for periodic Jacobi matrices. In *Proceedings of the Fourth Symposium on Basic Problems of Numerical Mathematics, Plzeň, Czechoslovakia 1978*, pages 63–76. Charles University, Prague, 1978. Cited in MR81e15005.

[7] D. L. Boley and G. H. Golub. The Lanczos–Arnoldi algorithm and controllability. *Systems and Control Letters*, 4:317–324, 1984.

[8] D. L. Boley and G. H. Golub. A modified method for reconstructing periodic Jacobi matrices. *Math. Comp.*, 42, 165:143–150, 1984.

[9] D. L. Boley and G. H. Golub. A survey of matrix inverse eigenvalue problems. *Inverse Problems*, 3:595–622, 1987.

[10] B. L. Buzbee, F. W. Dorr, J. A. George, and G. H. Golub. The direct solution of the discrete Poisson equation on irregular regions. *SIAM J. on Numerical Analysis*, 8:722–736, 1971.

[11] B. L. Buzbee, G. H. Golub, and C. W. Nielson. On direct methods for solving Poisson's equations. *SIAM J. on Numerical Analysis*, 7:627–656, 1970.

[12] A. K. Cline, G. H. Golub, and G. W. Platzman. Calculation of normal modes of oceans using a Lanczos method. In J. R. Bunch and D. J. Rose, editors, *Sparse Matrix Computations*, pages 409–426. Academic Press, New York, 1976.

[13] P. Comon and G. H. Golub. Tracking a few extreme singular values and vectors in signal processing. *IEEE Proceedings*, 78(8):1327–1343, 1990.

[14] P. Concus and G. H. Golub. Use of fast direct methods for the efficient numerical solution of nonseparable elliptic equations. *SIAM J. on Numerical Analysis*, 10:1103–1120, 1973.

[15] P. Concus and G. H. Golub. A generalized conjugate gradient method for nonsymmetric systems of linear equations. In R. Glowinski and J. L. Lions, editors, *Computing Methods in Applied Sciences and Engineering (2nd Internat. Symposium, Versailles 1975), Part I*, pages 56–65. Springer Lecture Notes in Econ. and Math. Systems 134, New York, 1976. Cited in MR57,7968.

[16] P. Concus, G. H. Golub, and G. Meurant. Block preconditioning for the conjugate gradient method. *SIAM J. Sci. Stat. Computing*, 6:220–252, 1985. Correction in 6(1985),p.791.

[17] P. Concus, G. H. Golub, and D. P. O'Leary. A generalized conjugate gradient method for the numerical solution of elliptic partial differential equations. In J. R. Bunch and D. J. Rose, editors, *Sparse Matrix Computations*, pages 309–332. Academic Press, New York, 1976. reprinted in *Studies in Numerical Analysis*, G. H. Golub (Ed.), Volume 25 of Studies in Mathematics, The Mathematical Association of America 1984 178-198.

[18] P. Concus, G. H. Golub, and D. P. O'Leary. Numerical solution of nonlinear elliptic partial differential equations by generalized conjugate gradient methods. *Computing*, 19:321–339, 1978.

[19] R. W. Cottle, G. H. Golub, and R. S. Sacher. On the solution of large structured linear complementarity problems: the block partitioned case. *Appl. Math. Optim.*, 4:347–363, 1978.

[20] G. Dahlquist, S. C. Eisenstat, and G. H. Golub. Bounds for the error of linear systems of equations using the theory of moments. *J. of Math. Anal. and Applics.*, 37:151–166, 1972.

[21] C. de Boor and G. H. Golub. The numerically stable reconstruction of a Jacobi matrix from spectral data. *Linear Alg. and Its Applics.*, 21:245–260, 1978.

[22] S. Elhay, G. H. Golub, and J. Kautsky. Updating and downdating of orthogonal polynomials with data fitting applications. In G.H. Golub and P. Van Dooren, editors, *Numerical Linear Algebra: Digital Signal Processing and Parallel Algorithms*, pages 149–172. Springer, New York, 1991. Citation unchecked.

[23] S. Elhay, G. H. Golub, and J. Kautsky. Updating and downdating of orthogonal polynomials with data fitting applications. *SIAM J. on Matrix Anal. and Applics.*, 12:327–353, 1991.

[24] S. Elhay, G. H. Golub, and J. Kautsky. Jacobi matrices for sums of weight functions. *BIT*, to appear, 1992.

[25] H. C. Elman and G. H. Golub. Iterative methods for cyclically reduced non-self-adjoint linear systems I. *Math. Comp.*, 54:671–700, 1989.

[26] H. C. Elman and G. H. Golub. Iterative methods for cyclically reduced non-self-adjoint linear systems II. *Math. Comp.*, 56:215–242, 1989.

[27] H. C. Elman and G. H. Golub. Block iterative methods for cyclically reduced nonselfadjoint elliptic problems. In D. R. Kincaid and L. J. Hayes, editors, *Iterative Methods for Large Linear Systems (Austin, TX 1988)*, pages 91–105. Academic Press, Boston, 1990. Cited in MR91f:65071.

[28] H. C. Elman and G. H. Golub. Line iterative methods for cyclically reduced non-self-adjoint elliptic problems. In *Transactions of the Seventh Army Conference on Applied Mathematics and Computing, West Point, NY, 1989.* ARO Report 90-1, US Army Research Office, Research Triangle Park, NC, 1990. Cited in MR90m00050.

[29] H. C. Elman and G. H. Golub. Line iterative methods for cyclically reduced discrete convection-diffusion problems. *SIAM J. Sci. and Stat. Comput.*, to appear, 1992.

[30] W. R. Ferng, G. H. Golub, and R. J. Plemmons. Adaptive Lanczos methods for recursive condition estimation. *Numerical Algorithms*, 1:1–19, 1991.

[31] J. H. Ferziger, G. H. Golub, and M. C. Thompson. Block SOR applied to the cyclically-reduced equations as an efficient solution technique for convection-diffusion equations. In *Computational Techniques and Applications CTAC-87.* Elsevier Science Publishers, New York, 1987. Citation unchecked.

[32] B. Fischer and G. H. Golub. On generating polynomials which are orthogonal over several intervals. *Math. Comp.*, to appear, 1992.

[33] R. W. Freund, G. H. Golub, and N. M. Nachtigal. Iterative solutions of linear systems. *Acta Numerica*, to appear, 1992.

[34] W. Gander, G. H. Golub, and D. Gruntz. Solving linear equations by extrapolation. In J. S. Kovalik, editor, *Supercomputing*, pages 279–293. Springer, 1989.

[35] G. H. Golub. The use of Chebyshev matrix polynomials in the iterative solution of linear equations compared with the method of successive over-relaxation. Technical report, Ph.D. Dissertation, University of Illinois, 1959.

[36] G. H. Golub. Bounds for the round-off errors in the Richardson second order method. *BIT*, 2:212–223, 1962.

[37] G. H. Golub. Some uses of the Lanczos algorithm in numerical linear algebra. In J. J. H. Miller, editor, *Topics in Numerical Analysis*, pages 173–184, New York, 1973. Academic Press.

[38] G. H. Golub. Bounds for matrix moments. *Rocky Mountain J. of Math*, 4:207–211, 1974.

[39] G. H. Golub. Methods for computing eigenvalues of sparse matrix equations. In V. Pereyra, editor, *Actas del Seminario Sobre Metodos Numericos Modernos, Volume I*, pages 125–148. Universidad Central de Venezuela, 1974. Citation unchecked.

[40] G. H. Golub. Sparse matrix computations: Eigenvalues and linear equations. In *Seminaries IRIA*, pages 117–140, Rocquencourt, 1975. Inst. Rech. D'Informatique et d'Automatique. Citation unchecked.

[41] G. H. Golub. Error bounds for iterative methods. Technical report, Numerical Analysis Project, Computer Science Department, Stanford University, Stanford, California, 1985.

[42] G. H. Golub and J. E. de Pillis. Toward an effective two-parameter SOR method. In D. R. Kincaid and L. J. Hayes, editors, *Iterative Methods for Large Linear Systems (Austin, TX 1988)*, pages 107–119. Academic Press, Boston, 1990. Cited in MR91f:65071.

[43] G. H. Golub and M. H. Gutknecht. Some modified moments for indefinite weight functions. *Numer. Math.*, 57:607–624, 1989.

[44] G. H. Golub, M. Heath, and G. Wahba. Generalized cross-validation as a method for choosing a good ridge parameter. *Technometrics*, 21(2), 1979.

[45] G. H. Golub and W. Kahan. Calculating the singular values and pseudo-inverse of a matrix. *SIAM J. on Numerical Analysis*, 2 (Series B):205–224, 1965.

[46] G. H. Golub and R. Kannan. Convergence of a two-stage Richardson process for nonlinear equations. *BIT*, 26:209–216, 1986.

[47] G. H. Golub and J. Kautsky. Calculation of Gauss quadratures with multiple free and fixed knots. *Numer. Math.*, 41:147–163, 1983.

[48] G. H. Golub and J. Kautsky. On the calculation of Jacobi matrices. *Linear Alg. and Its Applics.*, 52:439–455, 1983.

[49] G. H. Golub and M. Kent. Estimation of eigenvalues for iterative methods. *Math. Comp.*, 53:619–626, 1989.

[50] G. H. Golub, F. T. Luk, and M. L. Overton. A block Lanczos method for computing the singular values and corresponding singular vectors of a matrix. *ACM Trans. Math. Software*, 7:149–169, 1981.

[51] G. H. Golub, P. Manneback, and Ph. L. Toint. A comparison between some direct and iterative methods for large scale geodetic least squares problems. *SIAM J. Sci. Stat. Computing*, 7:799–816, 1986.

[52] G. H. Golub and D. Mayers. The use of pre-conditioning over irregular regions. In R. Glowinski and J. L. Lions, editors, *Computing Methods in Applied Science and Engineering VI*, pages 3–14. Elsevier Science Pub. Co., New York, 1984.

[53] G. H. Golub and G. A. Meurant. *Résolution Numérique Des Grands Systèmes Linéaires (Numerical Solution of Large Linear Systems)*. Collection de la Direction des Études et Recherches d'Electricité de France 49, Éditions Eyrolles, Paris, 1983. 329 pages. Cited in MR86b:65002.

[54] G. H. Golub and S. Nash. Nonorthogonal analysis of variance using a generalized conjugate gradient algorithm. *J. Amer. Statis. Assoc.*, 77:109–116, 1982.

[55] G. H. Golub and D. P. O'Leary. Some history of the conjugate gradient and Lanczos algorithms: 1948-1976. *SIAM Review*, 31:50–102, 1989.

[56] G. H. Golub and M. L. Overton. Convergence of a two-stage Richardson iterative procedure for solving systems of linear equations. In G. A. Watson, editor, *Numerical Analysis Proceedings, Dundee 1981*, pages 125–139. Springer Lecture Notes in Math. 912, New York, 1981.

[57] G. H. Golub and M. L. Overton. The convergence of inexact Chebyshev and Richardson iterative methods for solving linear systems. *Numer. Math.*, 53:571–593, 1988.

[58] G. H. Golub and V. Pereyra. The differentiation of pseudo-inverses and nonlinear least squares problems whose variables separate. *SIAM J. on Numerical Analysis*, 10:413–432, 1973.

[59] G. H. Golub and V. Pereyra. Differentiation of pseudoinverses, separable nonlinear

least squares problems and other tales. In M. Z. Nashed, editor, *Generalized Inverses and Applications*, pages 303–324, New York, 1976. Academic Press.

[60] G. H. Golub, R. J. Plemmons, and A. Sameh. Parallel block schemes for large-scale least-squares computations. In *High-Speed Computing Scientific Applications and Algorithm Design*, pages 171–179. University of Illinois Press, Urbana, Illinois, 1988. Citation unchecked.

[61] G. H. Golub and T. N. Robertson. A generalized Bairstow algorithm. *Commun. of the ACM*, 10:371–373, 1967.

[62] G. H. Golub and E. Seneta. Computation of the stationary distribution of an infinite Markov matrix. *Bull. Austral. Math. Soc.*, 8:333–341, 1973.

[63] G. H. Golub and R. Underwood. The block Lanczos method for computing eigenvalues. In J. R. Rice, editor, *Mathematical Software III*, pages 361–377. Academic Press, New York, 1977.

[64] G. H. Golub, R. Underwood, and J. H. Wilkinson. The Lanczos algorithm for the symmetric $Ax = \lambda Bx$ problem. Technical Report STAN-CS-72-270, Stanford University Computer Science Department Report, Stanford, California, 1972.

[65] G. H. Golub and C. F. Van Loan. *Matrix Computations*. Johns Hopkins University Press, Baltimore, Maryland, 1983.

[66] G. H. Golub and C. F. Van Loan. *Matrix Computations*. Johns Hopkins University Press, Baltimore, Maryland, 2nd edition, 1989.

[67] G. H. Golub and R. S. Varga. Chebyshev semi-iterative methods, successive overrelaxation iterative methods, and second order Richardson iterative methods. *Numer. Math.*, 3:147–168, 1961.

[68] G. H. Golub and J. H. Wilkinson. Iterative refinement of least square solution. In W. A. Kalenich, editor, *Proceedings of the IFIP Congress 65, Vol. 2*, pages 606–607, Washington, D.C., 1966. Spartan Books.

[69] G. H. Golub and J. H. Wilkinson. Iterative refinement of least squares solution. *Numer. Math.*, 9:139–148, 1966.

[70] M. Pagano, G. H. Golub, and F. T. Luk. A large sparse least squares problem in photogrammetry. In J. F. Gentleman, editor, *Proceedings of the Computer Science and Statistics: 12th Annual Symposium on the Interface*, pages 26–28. University of Waterloo, Waterloo, Ontario, 1979.

# COMPUTING THE SPARSE SINGULAR VALUE DECOMPOSITION VIA SVDPACK

MICHAEL W. BERRY *

**Abstract.** SVDPACK comprises four numerical (iterative) methods for computing the singular value decomposition (SVD) of large sparse matrices using double precision ANSI Fortran-77. This software package implements Lanczos and subspace iteration-based methods for determining several of the largest singular triplets (singular values and corresponding left- and right-singular vectors) for large sparse matrices. The package has been ported to a variety of machines ranging from supercomputers to workstations: CRAY Y-MP, CRAY-2S, Alliant FX/80, IBM RS/6000-550, DEC 5000-100, and Sun SPARCstation 2. The performance of SVDPACK as measured by its use in computing large rank approximations to sparse *term-document* matrices from information retrieval applications, and on synthetically-generated matrices having clustered and multiple singular values is presented.

**1. Introduction.** The singular value decomposition (SVD) is commonly used in the solution of unconstrained linear least squares problems, matrix rank estimation, and canonical correlation analysis. In applications such as information retrieval, seismic reflection tomography, and real-time signal processing, the solution to these problems is needed in the shortest possible time. Given the growing availability of high performance computer systems, there has been great interest in the development of efficient implementations of the singular value decomposition, in general. In applications such as information retrieval ([5], [4]), the data matrix whose SVD is sought is usually large and sparse. It is this particular case that motivates SVDPACK, i.e., the implementation of SVD methods which can be used to determine singular values and singular vectors of large sparse matrices on a variety of machines. SVDPACK uses Lanczos, block-Lanczos, subspace iteration, and trace minimization methods for determining several of the largest singular values and corresponding singular vectors for *unstructured* sparse matrices arising from practical applications. Before discussing specific SVDPACK routines, we make a few definitions, and review a few of the fundamental characterizations of the SVD.

Without loss of generality, suppose $A$ is a sparse $m$ by $n$ ($m \gg n$) matrix wi$(A) = r$. The singular value decomposition (SVD) of $A$ can be defined as

$$A = U \Sigma V^T, \tag{1.1}$$

* Department of Computer Science, University of Tennessee, 107 Ayres Hall, Knoxville, TN, 37996-1301 (berry@cs.utk.edu). This research was supported by the National Science Foundation under grant NSF CDA-9115428, Apple Computer Inc., Cupertino, CA, under contract C24-9100120, and by the Institute for Mathematics and its Applications with funds provided by the National Science Foundation.

where $U^TU = V^TV = I_n$ and $\Sigma = diag(\sigma_1, ..., \sigma_n)$, $\sigma_i > 0$ for $1 \le i \le r$, $\sigma_i = 0$ for $i \ge r+1$. The first $r$ columns of the orthogonal matrices $U$ and $V$ define the orthonormalized eigenvectors associated with the $r$ nonzero eigenvalues of $AA^T$ and $A^TA$, respectively. The singular values of $A$ are defined as the diagonal elements of $\Sigma$ which are the nonnegative square roots of the $n$ eigenvalues of $AA^T$. The set $\{u_i, \sigma_i, v_i\}$ is called the $i$-th singular triplet. The singular vectors (triplets) corresponding to large (small) singular values are called large (small) singular vectors (triplets). The development of SVDPACK was primarily motivated by the following problem:

*Given the sparse $m \times n$ matrix $A$ and $p < n$, determine the p-largest singular triplets of $A$ as defined by* (1.1).

To illustrate ways in which the SVD can reveal important information about the structure of a matrix we state two well-known theorems:

THEOREM 1.1. *Let the SVD of $A$ be given by (1.1) and*

$$\sigma_1 \ge \sigma_2 \ge \cdots \ge \sigma_r > \sigma_{r+1} = \cdots = \sigma_n = 0 \ ,$$

*and let $R(A)$ and $N(A)$ denote the range and null space of $A$, respectively, then*

1. *Rank property: $rank(A) = r$, $N(A) \equiv span\{v_{r+1}, \cdots, v_n\}$, and $R(A) \equiv span\{u_1, \cdots, u_r\}$, where $U = [u_1 \; u_2 \; \cdots \; u_m]$ and $V = [v_1 \; v_2 \; \cdots \; v_n]$.*
2. *Dyadic decomposition: $A = \sum_{i=1}^{r} u_i \cdot \sigma_i \cdot v_i^T$ .*
3. *Norms: $\|A\|_F^2 = \sigma_1^2 + \cdots + \sigma_r^2$, and $\|A\|_2^2 = \sigma_1$.*

The rank property, perhaps one of the most valuable aspects of the SVD, allows us to use the singular values of $A$ as quantitative measures of the qualitative notion of rank. The dyadic decomposition, which is the rationale for data reduction or compression in many applications, provides a canonical description of a matrix as a sum of $r$ rank-one matrices of decreasing importance, as measured by the singular values. The three results in Theorem 1.1 can be combined to yield the following quantification of matrix rank deficiency (see [7] for a proof):

THEOREM 1.2. *[Eckart and Young] Let the SVD of $A$ be given by (1.1) with $r = rank(A) \le p = min(m,n)$ and define:*

$$A_k = \sum_{i=1}^{k} u_i \cdot \sigma_i \cdot v_i^T \ \textit{with} \ k < r \ ,$$

*then*

$$\min_{r(B)=k} \|A - B\|_F^2 = \|A - A_k\|_F^2 = \sigma_{k+1}^2 + \cdots + \sigma_p^2 \ .$$

This important result, which indicates that $A_k$ is the best rank-$k$ approximation (in a least squares sense) to the matrix $A$, is the basis for concepts such as *data reduction* and *image enhancement*. In fact, $A_k$ is the best approximation to $A$ for any unitarily invariant norm ([8]). Hence,

$$\min_{r(B)=k} ||A - B||_2 = ||A - A_k||_2 = \sigma_{k+1}.$$

In the next section, we illustrate the applicability of Theorem 1.2 to problems in information retrieval which motivated the development of SVDPACK. In Section 3, we briefly compare the memory requirements and parameter choices for the pairs of Lanczos-based routines, (LAS1, LAS2) and (BLS1, BLS2), and subspace iteration-based pairs, (SIS1, SIS2) and (TMS1, TMS2). Details of each iterative method implemented in SVDPACK can be found in [1] and [2]. In Section 4, we provide performance results of SVDPACK on various high-performance computer systems using a test suite of sparse matrices collected from researchers in information retrieval. In addition, we compare the performance of the SVDPACK methods for resolving clustered or multiple singular values using diagonal test matrices. Details of future SVDPACK enhancements are given in Section 5.

**2. Applications.** Sparse linear least squares problems naturally arise in many *real*-world applications. The use of the sparse SVD to solve such problems is of current interest to researchers in fields such as query-based information retrieval and seismic reflection tomography. In this section we will briefly focus on the use of SVDPACK relative to information retrieval models. See [2] for a discussion of the role SVD plays in seismic tomography applications.

In [4] and [5] a new approach to automatic indexing and retrieval is discussed. It is designed to overcome a fundamental problem that plagues existing information retrieval techniques that try to match words of queries with words of documents. The problem is that users want to retrieve on the basis of conceptual topic or meaning of a document. There are usually many ways to express a given concept (*synonymy*), so the literal terms in a user's query may not match those of a relevant document. In addition, most words have multiple meanings (*polysemy*), so terms in a user's query will literally match terms in irrelevant documents.

The proposed *latent semantic indexing* (LSI) approach tries to overcome the problems of word-based access by treating the observed word to text-object association data as an unreliable estimate of the true, larger pool of words that could have been associated with each object. It is assumed there is some underlying latent semantic structure[1] in word usage

[1] *Semantic structure* refers to the correlation structure in the way in which individual words appear in documents; *semantic* implies only the fact that terms in a document may be taken as referents to the document itself or to its topic.

data that is partially obscured by the variability of word choice. Using the SVD defined in (1.1), we can estimate this latent structure and remove the obscuring *noise*.

Specifically, for an $m \times n$ *term-document* matrix $A$ whose $m$ rows and $n$ columns ($m \gg n$) correspond to terms and documents, respectively, we seek the closest (in a least squares sense) rank-$k$ ($k \ll n$) matrix

$$A_k = \sum_{i=1}^{k} u_i \cdot \sigma_i \cdot v_i^T \;\; with \; k < r \, , \tag{2.1}$$

given by Theorem 1.2. The idea is that the matrix $A_k$ captures the major associational structure in the matrix and removes the noise. Since relatively few terms are used as referents to a given document, the rectangular matrix $A = [a_{ij}]$ is quite sparse. The matrix element $a_{ij}$ indicates the frequency in which term $i$ occurs in document $j$. As discussed in [6], each *raw* term frequency is usually modified using a sophisticated weighting scheme (e.g., entropy weighting) which takes into account the distribution of terms over documents. Hence, the matrix element $a_{ij}$ may be either an integer or a rational number. Depending upon the size of the database from which the term-document is generated, the matrix $A$ can have several thousand rows and slightly fewer columns. Table 2.1 lists a few statistics of ten sparse term-document matrices that have been generated. We note that $\mu_r$ and $\mu_c$ are the average number of nonzeros per row and column, respectively. The *Density* of each sparse matrix listed in Table 2.1 is defined to be the ratio (Rows $\times$ Columns) / (Nonzeros).

TABLE 2.1
*Sparse Term-Document Matrix Specifications.*

| Data | Source | Columns | Rows | Nonzeros | Density | $\mu_c$ | $\mu_r$ |
|---|---|---|---|---|---|---|---|
| CISI | Bellcore | 1460 | 5143 | 66340 | 0.88 | 45.4 | 12.9 |
| CRAN | Bellcore | 1400 | 4997 | 78942 | 1.10 | 56.4 | 15.8 |
| MED | Bellcore | 1033 | 5831 | 52012 | 0.86 | 50.4 | 8.9 |
| MAG | Bellcore | 425 | 10337 | 80888 | 1.80 | 190.3 | 7.8 |
| TECH | Bellcore | 6535 | 16637 | 327244 | 0.30 | 50.0 | 20.0 |
| NEWS | Bellcore | 19660 | 35796 | 1879480 | 0.02 | 95.6 | 50.0 |
| ENCY | Bellcore | 25629 | 56530 | 2843956 | 0.002 | 110.9 | 50.3 |

By using the *reduced model* in (2.1), usually with $k = \alpha n$ ($\alpha \leq .01$), minor differences in terminology are virtually ignored. Moreover, the closeness of objects is determined by the overall pattern of term usage, so documents can be classified together regardless of the precise words that are used to describe them, and their description depends on a consensus of their term meanings, thus dampening the effects of polysemy. As a result, terms that do not actually appear in a document may still be used as referents, if that is consistent with the major patterns of association in the data. Position in the reduced space ($R(A_k)$) then serves as a new kind of *semantic indexing*.

As discussed in [1] and [4], LSI using the sparse SVD can be more robust and economical than straight term overlap methods. However, in practice, one must compute at least 100-200 largest singular values and corresponding singular vectors of sparse matrices having similar characteristics to those matrices in Table 2.1. In addition, it is not necessarily the case that $\text{rank}(A) = n$ for the $m \times n$ term-document matrix $A$, this is due to errors caused by term extraction, spelling, or duplication of documents. Regarding the numerical precision of the desired singular triplets for LSI, recent tests using a few of the databases listed in Table 2.1 have revealed that the $i$-th residual, $\tilde{r}_i$, corresponding to the $i$-th approximate singular triplet, $\{\tilde{u}_i, \tilde{\sigma}_i, \tilde{v}_i\}$, need only satisfy

$$10^{-6} \leq ||\tilde{r}_i||_2 \leq 10^{-3},$$

where $||\tilde{r}_i||_2$ is defined by

$$||\tilde{r}_i||_2 = \left[ (||A\tilde{v}_i - \tilde{\sigma}_i \tilde{u}_i||_2^2 + ||A^T \tilde{u}_i - \tilde{\sigma}_i \tilde{v}_i||_2^2)^{\frac{1}{2}} \right] / \left[ ||\tilde{u}_i||_2^2 + ||\tilde{v}_i||_2^2 \right]^{\frac{1}{2}} .$$

Finally, as the desire for using LSI on larger and larger databases or archives grows, fast algorithms for computing the sparse singular value decomposition will become of paramount importance.

## 3. Algorithms.

There are two canonical sparse symmetric eigenvalue problems which can be used to (indirectly) compute the sparse singular value decomposition. In this section, we present various iterative methods which can be applied to these sparse symmetric eigenvalue problems.

### 3.1. Equivalent eigenvalue problems.

Associated with an $m \times n$ $(m \geq n)$ matrix $A$ is the symmetric $(m+n) \times (m+n)$ matrix

$$B = \begin{pmatrix} O & A \\ A^T & O \end{pmatrix} . \tag{3.1}$$

If $\text{rank}(A) = n$, it can be easily shown that the eigenvalues of $B$ are the $n$ pairs, $\pm\sigma_i$, where $\sigma_i$ is a singular value of $A$, with $(m-n)$ additional zero eigenvalues if $m > n$. The multiplicity of the zero eigenvalue of $B$ is $m + n - 2r$, where $r$=rank($A$). As shown in [2], if $(u_i, v_i)^T$ denotes an eigenvector of norm $\sqrt{2}$ corresponding to any positive eigenvalue, $\sigma_i$, of $B$, then $\sigma_i$ is a singular value of $A$ and $u_i$, $v_i$ are respectively, left and right singular vectors of $A$ corresponding to $\sigma_i$.

The numerical accuracy of the $i$-th approximate singular triplet $(\tilde{u}_i, \tilde{\sigma}_i, \tilde{v}_i)$ as determined via the eigensystem of the *2-cyclic*[2] matrix $B$ (provided

[2] A non-negative irreducible matrix $B$ which is 2-cyclic has 2 eigenvalues of modulus $\rho(B)$, where $\rho(B)$ is the spectral radius of $B$. See Definition 2.2 on page 35 in [9].

$A \geq 0$) is then determined by the norm of the eigenpair residual vector $r_i$ defined as

$$\|r_i\|_2 = \left[\|B(\tilde{u}_i, \tilde{v}_i)^T - \tilde{\sigma}_i(\tilde{u}_i, \tilde{v}_i)^T\|_2\right] / \left[\|\tilde{u}_i\|_2^2 + \|\tilde{v}_i\|_2^2\right]^{\frac{1}{2}} ,$$

which can also be written as

$$(3.2)\ \|r_i\|_2 = \left[(\|A\tilde{v}_i - \tilde{\sigma}_i\tilde{u}_i\|_2^2 + \|A^T\tilde{u}_i - \tilde{\sigma}_i\tilde{v}_i\|_2^2)^{\frac{1}{2}}\right] / \left[\|\tilde{u}_i\|_2^2 + \|\tilde{v}_i\|_2^2\right]^{\frac{1}{2}} .$$

Alternatively, we may compute the SVD of $A$ indirectly by the eigenpairs of either the $n \times n$ matrix $A^TA$ or the $m \times m$ matrix $AA^T$. If $V = \{v_1, v_2, ..., v_r\}$ are linearly independent $n \times 1$ eigenvectors of $A^TA$ so that $V^T(A^TA)V = \text{diag}(\sigma_1^2, \sigma_2^2, ..., \sigma_r^2, 0, ..., 0)$, then $\sigma_i$ is the $i$-th nonzero singular value of $A$ corresponding to the right singular vector $v_i$. The corresponding left singular vector, $u_i$, is then obtained as $u_i = \frac{1}{\sigma_i}Av_i$. Similarly, if $U = \{u_1, u_2, ..., u_r\}$ are linearly independent $m \times 1$ eigenvectors of $AA^T$ so that $U^T(AA^T)U = \text{diag}(\sigma_1^2, \sigma_2^2, ..., \sigma_r^2, 0, ..., 0)$, then $\sigma_i$ is the $i$-th nonzero singular value of $A$ corresponding to the left singular vector $u_i$. The corresponding right singular vector, $v_i$, is then obtained as $v_i = \frac{1}{\sigma_i}A^Tu_i$.

Computing the SVD of $A$ via the eigensystems of either $A^TA$ or $AA^T$ may be adequate for determining several of the largest singular triplets of $A$, but the loss of accuracy can be severe for the smallest singular triplets (see [3]). Whereas the smallest and largest singular values of $A$ are the extremes of the spectrum of $A^TA$ or $AA^T$, the smallest singular values of $A$ lie at the center of the spectrum of $B$ in (3.1). For computed eigenpairs of $A^TA$ and $AA^T$, the norms of the $i$-th eigenpair residuals (corresponding to (3.2)) are given by

$$\|r_i\|_2 = \|A^TA\tilde{v}_i - \tilde{\sigma}_i^2\tilde{v}_i\|_2 \ / \ \|\tilde{v}_i\|_2$$

and

$$\|r_i\|_2 = \|AA^T\tilde{u}_i - \tilde{\sigma}_i^2\tilde{u}_i\|_2 \ / \ \|\tilde{u}_i\|_2 ,$$

respectively. Thus, extremely high precision in computed eigenpairs may be necessary to compute the smallest singular triplets of $A$. Difficulties in approximating the smallest singular values by any of the three equivalent symmetric eigenvalue problems are discussed in [1]. The naming convention of each SVDPACK program specfies both the algorithm and the type of equivalent eigensystem used to approximate singular triplets. Specifically, all possible entries for the three fields of the four character SVDPACK filename `MMTE` are given in Table 3.1.

**4. Comparison of SVDPACK methods.** In this section, we briefly compare the memory requirements and parameter selection of the iterative methods implemented in SVDPACK. For a more detailed comparison of these methods (complexity, parallelism, and convergence properties) see [1] and [2].

TABLE 3.1
*SVDPACK program naming convention.*

| MMTE | | |
|---|---|---|
| Field | Description | Possible Entries |
| MM | Algorithm | BL ≡ Block Lanczos<br>LA ≡ Single Vector Lanczos<br>SI ≡ Subspace Iteration<br>TM ≡ Trace Minimization |
| T | File Type | D ≡ Documentation File<br>I ≡ Input File<br>O ≡ Output File<br>S ≡ Source File |
| E | Eigensystem<br>or<br>I/O Channel | 1 ≡ Cyclic Matrix $B$ defined by (3.1)<br>2 ≡ $A^TA$ Matrix<br>1,5,7 ≡ Input Channel<br>2,3,8,9 ≡ Output Channel |

**4.1. Memory.** Although the effective CPU time required by SVDPACK methods is certainly an essential metric in measuring their performance within applications such as information retrieval, it is also important to assess the amount of computer memory that is required. We estimate the amount of memory in 64-bit words (8 bytes) which is required by the four methods when equivalent eigensystems of the 2-cyclic matrix $B$ and $A^TA$ are used to determine the $p$-largest singular triplets of an $m \times n$ sparse matrix $A$. The estimates are based on our implementations of each method on both a Cray-2S/4-128[3] (static memory allocation only) and an Alliant FX/80[4] (static and dynamic memory allocation). We exclude the memory needed to store the matrix $A$ in the Harwell/Boeing sparse matrix format in these counts. Also, we exclude the memory ($mp$ words) needed to compute the left singular vectors from the eigensystem of $A^TA$. The parameters for BLS1 and BLS2 include the initial block size, $b$, an upper bound on the dimension of the Krylov subspace, $c$. For LAS1 and LAS2, we also include a similar upper bound, $q$, for the order of the symmetric tridiagonal matrix generated. The parameter $s$ for SIS1(2) and TMS1(2) defines the number of vectors ($s \geq p$) for the initial subspace.

Table 4.1 lists the required megabytes ($8 \times 10^{-6} \times$ (*number of words*)) for the SVDPACK methods when we require $p = 100$ of the largest singular triplets of selected sparse matrices from Table 1. The other parameters used

[3] CRAY-2S/4-128 located at the National Center for Supercomputing Applications (NCSA) at the University of Illinois at Urbana-Champaign.

[4] Alliant FX/80 located at the Center for Supercomputing Research and Development (CSRD) at the University of Illinois at Urbana-Champaign.

for these memory counts are

$$b = 30,\ c = 150,\ s = 120,\ q = \min\{\tilde{n}, 600\}\ ,$$

where $\tilde{n} = n$ when approximating eigensystems of $A^T A$, and $\tilde{n} = m + n$ when approximating the eigensystem of $B$. Irrespective of which eigensystems are approximated ($B$ or $A^T A$), the term-document matrix **TECH** will pose the greatest memory constraint among those listed in Table 4.1. Since we must have $s \geq 2p$ for SIS1 and LAS1 when approximating eigenpairs of $B$ (see [2]), the entries in Table 4.1 reflect the selection of $s = 200$ for SIS1 and $p = 200$ for LAS1 .

With regard to memory consumption, Table 4.1 reveals that SIS2 is the most *conservative* of the four methods while LAS1 can be considered the most *liberal.* The memory requirement of LAS1 is nearly 5 times that of SIS1 across most of the term-document matrices. When the eigensystem of $A^T A$ is approximated, however, the factor is approximately 3. For most of the matrices in Table 1, the savings in memory for all SVDPACK methods ranges from a factor of 1.5 to 5 when we approximate eigenpairs of $A^T A$ rather than eigenpairs of the order $m + n$ matrix $B$. However, for suitably rectangular matrices ($m \gg n$) such as the $10337 \times 425$ matrix **TIME**, the savings can be enormous. Specifically, LAS2 and TMS2 each require a factor 22 fewer megabytes whereas the factors for SIS2 and BLS2 are 17 and 19, respectively. Hence, we can have an important trade-off in memory versus accuracy when we choose to determine singular triplets of the matrix $A$ via the eigensystem of $A^T A$. In approximating the $p$-largest triplets of $A$, the effect of scaling for the left singular vector on the residual in (3.2) will determine whether or not the less-stringent memory requirement associated with $A^T A$ is a mixed blessing.

TABLE 4.1
*Memory (in megabytes) required by SVDPACK methods when computing the 100-largest triplets of matrices in Table 1 via eigensystems of the 2-cyclic matrix $B$ and $A^T A$.*

| Matrix | Memory in Megabytes | | | | | | | |
|---|---|---|---|---|---|---|---|---|
| | LA– | | BL– | | SI– | | TR– | |
| | S1 | S2 | S1 | S2 | S1 | S2 | S1 | S2 |
| TIME | 55 | 4 | 31 | 3 | 21 | 1 | 27 | 2 |
| MED | 56 | 10 | 32 | 6 | 22 | 2 | 28 | 4 |
| CISI | 58 | 12 | 35 | 8 | 22 | 3 | 29 | 6 |
| CRAN | 90 | 12 | 57 | 8 | 35 | 3 | 45 | 6 |
| TECH | 153 | 45 | 79 | 37 | 60 | 14 | 79 | 28 |

**4.2. Parameter estimation.** It is important to note the difficulties (if any) in selecting *good* parameters for these iterative methods. In the subspace methods, SIS1(2) and TMS1(2), we seek the smallest initial subspace dimension $s$, where $s \geq p$, that can approximate the desired $p$-largest singular triplets of $A$. For TMS2, we must also choose an appropriate context switch parameter, $\eta$, to switch from polynomial acceleration to shifting by Ritz values (see [2]). We have been fairly successful with the choices $\eta = 10^0, 10^{-1}$, however, for most of the matrices in Table 1.

As mentioned in the previous section, in applying SIS1 with the indefinite 2-cyclic operator, $B$, it is best to choose $s$ to be at least twice the size ($2p$) of desired singular triplets. The same situation holds for LAS1 when we approximate the eigensystem of $B$.

BLS1 and BLS2 require an initial block size $b$, where $b \leq p$, and the bound $c$ on the Krylov subspace generated within the block Lanczos recursion. The choice of $b$ can be difficult, and as mentioned in [1] some gains have been made in removing this ambiguity by incorporating spectral estimation via Gershgorin disks. This block size adjustment strategy, however, is not necessarily *robust* and may warrant further refinement. The choice of $c$, however, is perhaps one parameter whose value is more or less determined by the computer system rather than the programmer. As with the Krylov bound, $q$, for LAS1 and LAS2 , we can certainly choose $c$ to be the order of the eigensystem used to approximate the desired singular triplets ($m + n$ for $B$ or $n$ for $A^T A$), but the available memory (in 64-bit words) is perhaps the best choice for $c$.

**5. Performance.** In this section, we present sample SVDPACK benchmarks on machines ranging from supercomputers to workstations. Model SVD problems using matrices from Tables 1 are solved using SVDPACK on the Alliant FX/80, CRAY-2S/4-128, and IBM RS/6000-550 computer systems. Due to memory limitations on some machines (e.g., Alliant FX/80 and IBM RS/6000-500), we do not have a complete set of benchmarks for all SVDPACK methods on each machine considered.

**5.1. Model problems.** In the following experiments, we determine 100 of the largest singular triplets of some of the matrices ($A$) in Table 1. We determine these triplets to residual norms (3.2) of order $10^{-3}$ (at least) via eigensystems of either $A^T A$ or the 2-cyclic matrix $B$. Plots of the desired singular values for selected matrices from Table 1 are provided in Figure 5.1.

As discussed in Section 4.2, with LAS1 we must request $k = 2p = 200$ singular triplets of the $m \times n$ sparse matrix $A$ if we are to determine the largest $p$ singular triplets by considering the eigensystem of the matrix $B$. The same situation is certainly true for SIS1, although we can also compute eigensystems of the positive definite matrix $\tilde{B} = B + \gamma I$, where $\gamma = ||A||_1$. Similarly, TMS1 and TMS2 can approximate eigensystems of $\tilde{B}$ and $\gamma^2 I_n - A^T A$, respectively, for these problems.

For our Lanczos-based methods, LAS1(2) and BLS1(2), we limit the dimension of Krylov subspaces by selecting

$$c = 150, \textit{ and } q = \min\{n, 600\},$$

for BLS1(2) and LAS1(2), respectively. Our initial block size for BLS1(2) is $b = 30$ and the initial subspace dimension for SIS1(2) and TMS1(2) is $s = 120$. For TMS2, we choose $\eta = 10^{-1}$ as our residual reduction factor for the context switch from polynomial acceleration to shifting by Ritz values.

Table 5.1 lists several SVDPACK benchmarks (CPU times in seconds) on the CRAY-2S/4-128. We observe that BLS1 can be between 1.2 to 1.8 times faster than LAS1 for most problems (the exception being **TECH** for which both times are comparable). Requiring $k = 2p = 200$ singular triplets to be approximated by LAS1 practically *doubles* the number of Lanczos steps normally required to approximate $k = 100$ eigenvalues. This requirement, of course, is not necessary if eigensystems of $A^T A$ can be used to approximate these triplets. We especially note the superiority of LAS2 over BLS2 when we approximate the 100-largest eigenpairs of $A^T A$ as opposed to those of $B$. Specifically, LAS2 ranges between 2.6 to 6.7 times faster than BLS2 across our suite of matrices. Hence, when approximating eigensystems of $B$ and $A^T A$, we have observed that using LAS2 to approximate the eigenpairs of $A^T A$ is the Lanczos scheme of choice for obtaining the largest singular triplets.

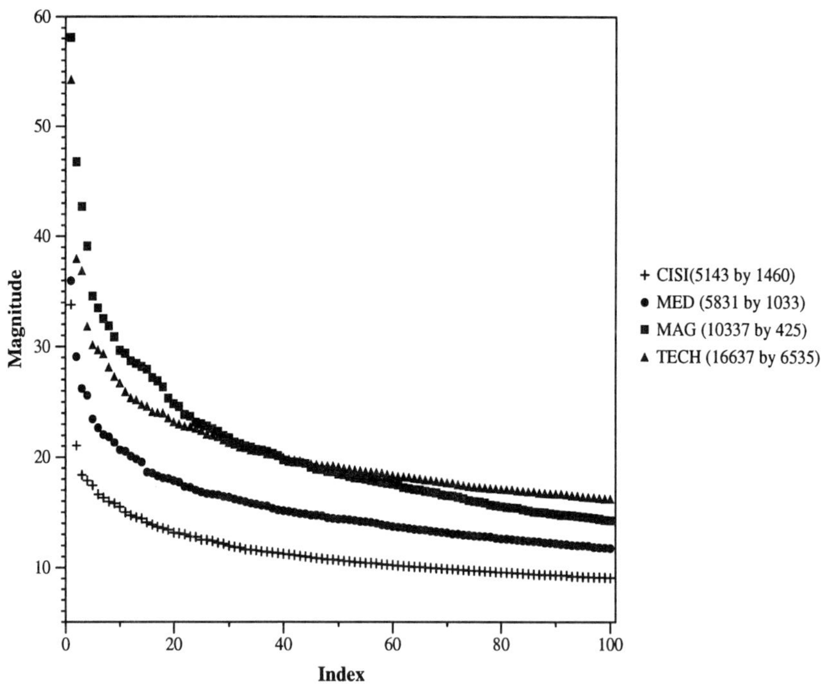


FIG. 5.1. *The* 100*-largest singular values of CISI, MAG, MED, and TECH from Table* 1.

To assess the specific gains in performance (time reduction) associated with the eigenvalue problem of order $n$, we list the speed improvements for

LAS2 and BLS2 (and the other SVDPACK methods) when eigensystems of $A^T A$ are approximated in Table 5.2. The limited improvement for BLS2, in this case, stems from the fact that although the less time is spent in re-orthogonalization (see [2]) the number of outer iteration steps for the $A^T A$-based recursion can be as much as 1.5 times greater than the number of outer iterations for the cyclic-based hybrid recursion (see [2]). Hence, the deflation associated with *larger* gaps among the $p = 100$ eigenvalues of $A^T A$ is not quite as efficient. Also, the number of sparse matrix-vector multiplications per outer iteration step is 4 for the $A^T A$ implementation rather than 2 for the cyclic-based recursion.

Next, we compare the performance of the two versions of SIS1 with that of TMS1 when eigensystems of $B$ and $\tilde{B}$ are approximated. The effective speed improvement of SIS1($\tilde{B}$) over SIS1($B$) ranges from 1.08 to 3.98. This, in effect, is due to the length of the Chebyshev intervals, which certainly determines the polynomial of minimial degree needed to accelerate convergence. We note that Chebyshev polynomials of degrees as high as 44 are required by SIS1($B$), while TMS1 and SIS1($\tilde{B}$) require polynomials of maximum degree 2 and 22, respectively. As a result, TMS1 ranges between 2.2 to 4.96 times faster than SIS1($\tilde{B}$) for our test suite of matrices. Since, compiler-generated parallelism for TMS1, yielded an average of 1.68 active CPU's in a *non-dedicated* environment, it is reasonable to expect even higher speed improvements for a dedicated machine. When approximating eigenpairs of $A^T A$ on the CRAY-2S/4-128, we find SIS2 to be a bit more competitive with TMS2 being approximately 1.5 times faster across the suite of matrices. As indicated in Table 5.2, SIS2 is much more effective in approximating singular triplets of $A$ via eigenpairs of $A^T A$. For most matrices, SIS2 only required Chebyshev polynomials of degree 10. TMS2, on the other hand, required polynomials of degree 2 for each matrix. With the exception of BLS2, we note that the best performance on the Cray-2S/4-128 for our candidate methods is obtained when eigensystems of $A^T A$ or $\gamma^2 I_n - A^T A$ are approximated. From Table 5.1, we note that LAS2 ranges between 2.6 to 6.7 times faster than BLS2, and 1.9 to 4.3 times faster than TMS2 on the CRAY-2S/4-128.

On the Alliant FX/80, LAS2 is on average 2.7 times faster than BLS2, and 2.4 times faster than TMS2 (via eigensystems of $\gamma^2 I_n - A^T A$) The effective parallelization of TMS2 and SIS2 (see [2]) tends to produce comparable times for both methods (see Table 5.3) for the moderate-order matrices. For TMS2, the parallel conjugate gradient (CG) iterations can sustain an optimal performance rate (lower total execution time) with each processor handling one independent system.

**5.2. Clustered spectra.** In this section, we illustrate the performance of SVDPACK on diagonal test matrices having clustered or multiple singular values. On an IBM RS/6000-550 workstation, we used all 8 SVDPACK programs to compute the 10-largest singular triplets having

TABLE 5.1
*Cray-2S/4-128 CPU times (in seconds) for determining the 100-largest singular triplets of term-document matrices from Table 1.*

| SVD via eigensystems of $B$, $\tilde{B}$ | | | | | |
|---|---|---|---|---|---|
| Matrix | LAS1 | BLS1 | SIS1($B$) | SIS1($\tilde{B}$) | TMS1 |
| MED | 139 | 103 | 1269 | 333 | 136 |
| CISI | 143 | 120 | 1276 | 515 | 187 |
| TIME | 147 | 127 | 1616 | 634 | 300 |
| CRAN | 167 | 117 | 1105 | 874 | 176 |
| TECH | 479 | 486 | 5598 | 1405 | 636 |
| SVD via eigensystems of $A^TA$ | | | | | |
| Matrix | LAS2 | BLS2 | SIS2 | | TMS2 |
| MED | 16 | 86 | 88 | | 52 |
| CISI | 23 | 120 | 137 | | 76 |
| TIME | 13 | 87 | 93 | | 56 |
| CRAN | 22 | 117 | 120 | | 77 |
| TECH | 89 | 490 | 605 | | 292 |

residuals (3.2) of order $\mathcal{O}(10^{-6})$ or $\mathcal{O}(10^{-10})$. The spectra of these test matrices are specified by the function

$$\Sigma(\alpha,\ \beta,\ \delta,\ k_{max},\ i_{max}) = (\alpha k + \beta) + \delta i$$

where $k = 0, 1, \ldots, k_{max}$ is the index over the clusters, $i = 1, 2, \ldots, i_{max}$ is the index within a cluster, $\delta$ is the uniform separation between singular values within a cluster, and $\alpha$, $\beta$ are scalars. For example,

$$\Sigma(1, 0, 10^{-6}, 4, 10) = k + i \times 10^{-6},$$

TABLE 5.2
*Speed improvements for approximating the largest $p = 100$ singular triplets via eigensystems of $A^TA$ or $\gamma^2 I_n - A^TA$ as opposed to those of $B$ or $\tilde{B}$ (cyclic matrices) on the Cray-2S/4-128.*

| Matrix | $\frac{LAS1}{LAS2}$ | $\frac{BLS1}{BLS2}$ | $\frac{SIS1(\tilde{B})}{SIS2}$ | $\frac{TMS1}{TMS2}$ |
|---|---|---|---|---|
| CISI | 6.2 | 1.1 | 9.3 | 2.5 |
| CRAN | 11.3 | 1.5 | 17.3 | 5.4 |
| MED | 7.5 | 1.1 | 9.2 | 2.3 |
| TECH | 5.3 | 1.0 | 9.2 | 2.2 |
| TIME | 13.6 | 1.0 | 5.3 | 1.3 |

TABLE 5.3
*Alliant FX/80 CPU times (in seconds) for determining the 100-largest singular triplets of term-document matrices from Table 1.*

| SVD via eigensystems of $A^TA$ | | | | |
|---|---|---|---|---|
| Matrix | LAS2 | BLS2 | SIS2 | TMS2 |
| MED | 71 | 247 | 197 | 189 |
| CISI | 110 | 319 | 364 | 266 |
| TIME | 43 | 194 | 172 | 165 |
| CRAN | 95 | 250 | 315 | 277 |

where $k = 0, \ldots, 4$ and $i = 1, \ldots, 10$, specifies 5 clusters each of size 10 near 0, 1, 2, 3, and 4.

Four diagonal test matrices having clustered or multiple elements which have been used to test the robustness of the various SVDPACK methods are defined in Table 5.4. The number of clusters and separation of singular values within each cluster were deliberately chosen to reflect different clustering properties, as illustrated by the scatter plots of singular values in Figure 5.2. For matrices CLUS1 and CLUS2, we desire the 10-largest triplets having residuals of order $\mathcal{O}(10^{-6})$ or less, and for CLUS3 and CLUS4 we demand residuals of order $\mathcal{O}(10^{-10})$ or less.

TABLE 5.4
*Properties of the clustered spectra for the* $50 \times 50$ *diagonal matrices for SVDPACK testing.*

| Matrix | Number of Clusters | Cluster Separation ($\delta$) | Maximum Cluster Size | $\Sigma(\alpha, \beta, \delta, k_{max}, i_{max})$ |
|---|---|---|---|---|
| CLUS1 | 13 | $10^{-3}$ | 4 | $\Sigma(4, 0, \delta, 12, 4)$ |
| CLUS2 | 5 | $10^{-6}$ | 10 | $\Sigma(1, 0, \delta, 4, 10)$ |
| CLUS3 | 2 | $10^{-8}$ | 25 | $\Sigma(26, -\delta, \delta, 1, 25)$ |
| CLUS4 | 5 | 0 | 10 | $\Sigma(10, 1, \delta, 4, 10)$ |

With regard to parameter choices for LAS1(2) and BLS1(2), we limit the dimension of Krylov subspaces by selecting

$$c = 30, \ \textit{and} \ q = 50,$$

where $c$ is the maximum order of the block tridiagonal matrices generated by BLS1(2), and $q$ is the bound on the number of Lanczos steps in LAS1(2). Our initial block size for BLS1(2) is $b = 10$, and the initial subspace dimension for SIS1(2) and TMS1(2) is $s = 12$. For TMS2, we choose $\eta = 10^{-1}$ as our residual reduction factor for the context switch from polynomial acceleration to shifting by Ritz values. We do not employ polynomial acceleration within TMS2 for approximating the tightly-clustered and multiple singular values of matrices CLUS3 and CLUS4, respectively.

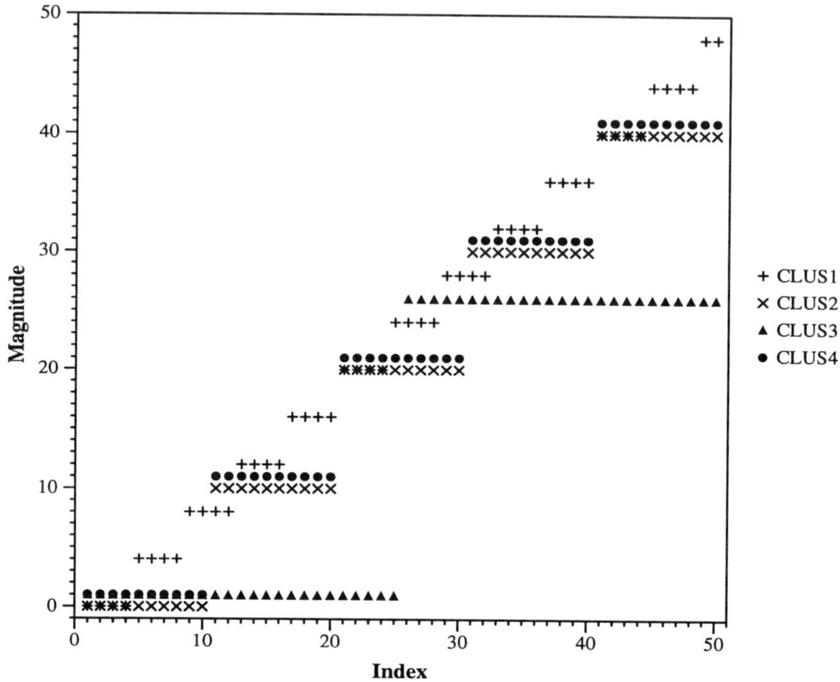


FIG. 5.2. *Clustered spectra of* $50 \times 50$ *test matrices.*

Figures 5.3 through 5.5 illustrate the effectiveness of each SVDPACK method for resolving clustered spectra with respect to efficiency and robustness. Whereas the block Lanczos (BLS1(2)) and trace minimization (TMS1(2)) methods consistently approximate the desired triplets, the single-vector Lanczos (LAS1(2)) and subspace iteration (SIS1(2)) methods are more erratic in that they may produce far more or less triplets than expected, depending on the clustering. As illustrated in Figure 5.3, LAS1 performs 100 iterations and determines all the singular values of `CLUS1` and `CLUS2`, whereas LAS2 and SIS1(2) can only resolve (to $\mathcal{O}(10^{-10})$ accuracy) a few of the poorly-separated singular values of the cluster near 26 for `CLUS3`. LAS1(2) are not reliable for estimating multiplicities (see also [3]) as evidenced by the multiplicity of 4 (as opposed to the true multiplicity of 10) for the singular value $\sigma = 41$ of matrix `CLUS4`.

Figure 5.4 illustrates the range of the error corresponding to the least-accurate singular triplet computed by each SVDPACK method. Here, we simply plot the smallest exponent (in absolute value) for the residual vectors defined by (3.2). Only BLS1(2) and TMS1(2), determined all the desired triplets at the specified accuracies for each test matrix. The use of polynomial acceleration in SIS1(2) fails to adequately dampen out the unwanted tightly-clustered singluar values of matrix `CLUS3`. The cost of using very high-degree polynomials, in this case, is the excessive number of sparse matrix multiplications (by $A$ and $A^T$) for SIS1(2) (see Figure 5.5). Chebyshev polynomials of degree 256 and higher were used by SIS1(2) to dampen-out the 15 unwanted singular values in the tight cluster of singular values near $\sigma = 26$. We report the total number of multiplications by the sparse matrix $A$ and its transpose for each method in Figure 5.5, rather than execution time on the IBM RS/6000-550, since the test matrices are diagonal and the cost of multiplication would be negligible. As expected, LAS2 is of lowest complexity when it adequately approximates the desired triplets (matrices CLUS1 and CLUS2), otherwise BLS1(2) and possibly TMS2 are more computationally viable.

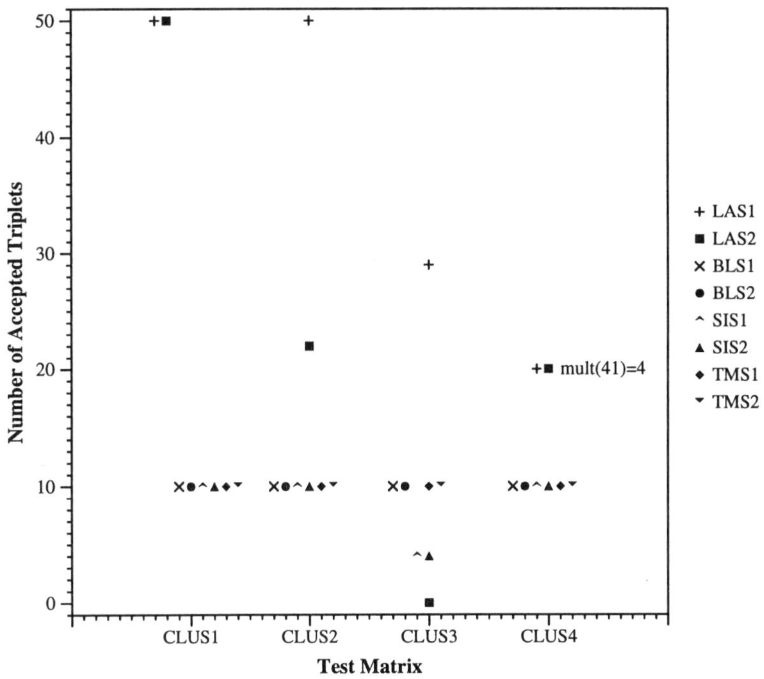


FIG. 5.3. *SVDPACK accuracy on* $50 \times 50$ *test matrices: number of acceptable triplets.*

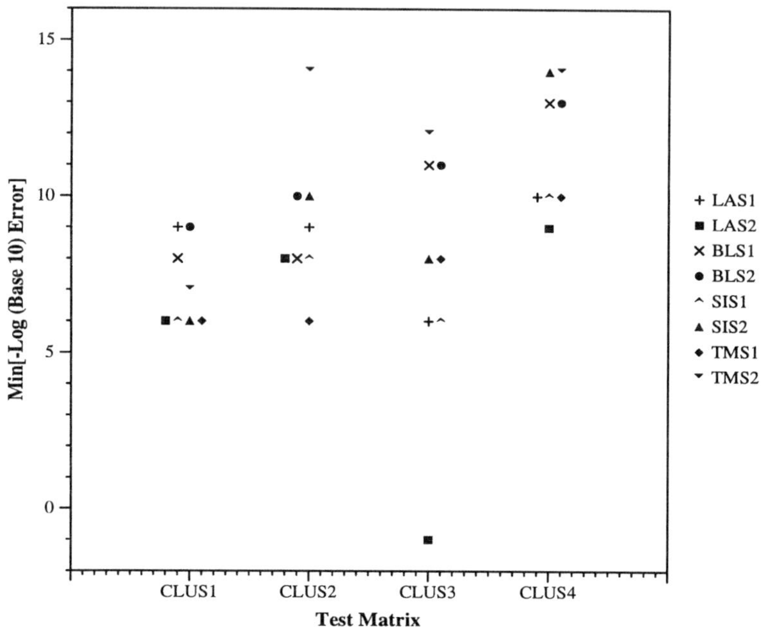


FIG. 5.4. *SVDPACK accuracy on* $50 \times 50$ *test matrices: norm of largest residual vector.*

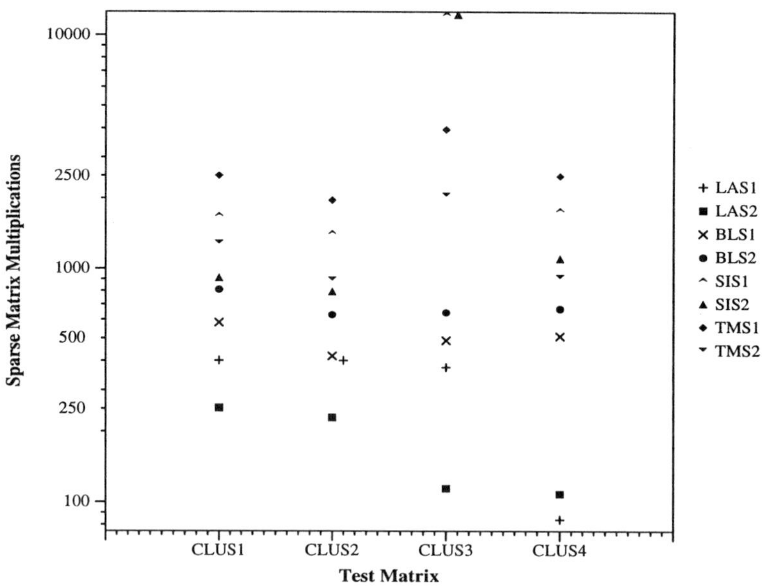


FIG. 5.5. *SVDPACK performance on* $50 \times 50$ *test matrices: number of sparse matrix multiplications.*

**6. Future work.** In addition to the continued porting of the Fortran-77 SVDPACK library to more machines, an effort is underway to produce an ANSI-C equivalent version of SVDPACK. For determining the singular value decomposition of extremely large sparse matrices using a modest computing environment, we anticipate the development of an out-of-core distributed version of SVDPACK which ports to a heterogeneous network of workstations. Implementations on massively-parallel computer systems such as the MasPar MP-1 and Thinking Machines CM-5 are planned as well.

The current Fortran-77 source code and documentation for SVDPACK may be obtained through the NETLIB facility maintained by the University of Tennessee and Oak Ridge National Laboratory. Users should send the electronic mail message *send index from svdpack* to `netlib@ornl.gov` to get a listing of the methods and associated files composing the library.

**7. Acknowledgements.** The author would like to thank Susan Dumais at Bell Communications Research (Bellcore), Morristown, NJ, and Dulce Ponceleon at Apple Computer Inc., Cupertino, CA, for their collaborative efforts in constructing a test suite of sparse matrices from information retrieval.

## REFERENCES

[1] M. W. BERRY, *Multiprocessor Sparse SVD Algorithms and Applications*, PhD thesis, University of Illinois at Urbana-Champaign, Urbana, IL, 1990.

[2] ———, *Large scale singular value computations*, International Journal of Supercomputer Applications, 6 (1992), pp. 13–49.

[3] J. K. CULLUM AND R. A. WILLOUGHBY, *Lanczos Algorithm for Large Symmetric Eigenvalue Computations, Volume 1 Theory*, Birkhäuser, Boston, 1985.

[4] S. DEERWESTER, S. DUMAIS, G. FURNAS, T. LANDAUER, AND R. HARSHMAN, *Indexing by latent semantic analysis*, Journal of the American Society for Information Science, 41 (1990), pp. 391–407.

[5] S. DUMAIS, G. FURNAS, AND T. LANDAUER, *Using latent semantic analysis to improve access to textual information*, in Proceedings of Computer Human Interaction '88, 1988.

[6] S. T. DUMAIS, *Improving the retrieval of information form external sources*, Behavior Research Methods, Instruments, & Computers, 23 (1991), pp. 229–236.

[7] G. GOLUB AND C. REINSCH, *Singular value decomposition and least squares solutions*, in Handbook for Automatic Computation II, Linear Algebra, Springer-Verlag, New York, 1971.

[8] L. MIRSKY, *Symmetric gage functions and unitarilly invariant norms*, Quarterly Journal of Mathematics, 11 (1960), pp. 50–59.

[9] R. S. VARGA, *Matrix Iterative Analysis*, Prentice Hall, Englewood Cliffs, NJ, 1962.

# GAUSSIAN QUADRATURE APPLIED TO ADAPTIVE CHEBYSHEV ITERATION

D. CALVETTI *, G.H. GOLUB †, AND L. REICHEL ‡

**Abstract.** Chebyshev iteration has been a popular iterative scheme for the solution of large linear systems of equations with a symmetric positive definite matrix $A$. With the advent of parallel processors, there has been a resurgence of interest in this method. In Chebyshev iteration one determines iteration parameters so that the residual polynomials are scaled Chebyshev polynomials for some interval $[a, b]$ on the positive real axis. Chebyshev iteration is often implemented as an adaptive iteration scheme, in which one during the iterations seeks to determine an interval $[a, b]$ that make the iterates converge rapidly to the solution of the linear system. Roughly, the interval should contain most of the spectrum of $A$ and be as small as possible. Recently, Golub and Kent [6] proposed a new adaptive Chebyshev iteration method, in which inner products of residual vectors are interpreted as modified moments. These modified moments and the recursion coefficients for the residual polynomials yield a symmetric tridiagonal matrix, whose eigenvalues are used to determine an interval $[a, b]$. The eigenvalues are nodes of a Gaussian quadrature rule. We propose a modification of this scheme, in which the determination of a suitable interval is based on the weights of this quadrature rule also. Computed examples illustrate that a significant reduction in the number of iterations can be achieved by this modification.

**1. Introduction.** We consider iterative solution of large linear systems of equations

$$Ax = b, \qquad A \in R^{n \times n}, \qquad x, b \in R^n, \tag{1.1}$$

with a symmetric and positive definite matrix. The Chebyshev iteration method is a classical iterative scheme for the solution of such systems; see [7,11]. It has received renewed attention recently because it is fairly simple to implement Chebyshev iteration efficiently on parallel and vector computers; see Dongarra et al. [2]. The attractiveness of carrying out Chebyshev iteration on multiprocessors stems from that no inner products are required when a suitable interval containing (most of) the spectrum of the matrix $A$ is known. In contrast, the conjugate gradient method requires two inner products per iteration.

Let $x_0$ be a given initial approximate solution of 1.1 and let $[a, b]$ be a given interval on the positive real axis. Chebyshev iteration with respect to the interval $[a, b]$ can be written in the form

$$x_{k+1} = x_{k-1} + \omega_{k+1}(\delta r_k + x_k - x_{k-1}), \qquad k = 0, 1, 2, \ldots, \tag{1.2}$$

* Department of Pure and Applied Mathematics, Stevens Institute of Technology, Hoboken, NJ 07030. e-mail: na.calvetti@na-net.ornl.gov Research supported in part by the Design and Manufacturing Institute at Stevens Institute of Technology.

† Department of Computer Science, Stanford University, Stanford, CA 94305. e-mail: na.golub@na-net.ornl.gov Research supported in part by NSF grant CCR-8821078.

‡ Department of Mathematics and Computer Science, Kent State University, Kent, OH 44242. e-mail: na.reichel@na-net.ornl.gov Research supported in part by NSF grants DMS-9002884 and DMS-9205531.

where $\boldsymbol{x}_{-1} = \boldsymbol{0}$ and $\boldsymbol{r}_k = \boldsymbol{b} - A\boldsymbol{x}_k$ is the residual vector associated with the iterate $\boldsymbol{x}_k$. Let

$$\delta = \frac{2}{b+a}, \qquad \mu = \frac{b-a}{b+a}.$$

Then the iteration parameters $\omega_{k+1}$ are defined by $\omega_1 = 1$ and

$$\omega_{k+1} = \frac{1}{1 - \frac{\mu^2}{4}\omega_k}, \qquad k = 1, 2, 3, \ldots .$$

The residual vectors $\boldsymbol{r}_k$ can be written as

$$\boldsymbol{r}_k = p_k(A)\boldsymbol{r}_0, \qquad k = 1, 2, 3, \ldots , \tag{1.3}$$

where the residual polynomials $p_k$ satisfy

$$p_k(\lambda) = \frac{C_k\left(\frac{\lambda'}{\mu}\right)}{C_k\left(\frac{1}{\mu}\right)}, \qquad \lambda' = 1 - \delta\lambda. \tag{1.4}$$

Here the $C_k$ are Chebyshev polynomials

$$C_k(\lambda) = \begin{cases} \cos(k\cos^{-1}\lambda), & |\lambda| \leq 1, \quad k \geq 0, \\ cosh(k\, cosh^{-1}\lambda), & |\lambda| > 1, \quad k \geq 0. \end{cases} \tag{1.5}$$

We refer to [7,9,11] for details. For future reference, we note that the residual polynomials satisfy the recurrence relation

$$\begin{aligned} \text{(1.6)}\ \lambda p_k(\lambda) &= \beta_k p_{k+1}(\lambda) + \alpha_k p_k(\lambda) + \gamma_k p_{k-1}(\lambda), \qquad k = 0, 1, 2, \ldots , \\ p_0(\lambda) &= 1, \quad p_{-1}(\lambda) = 0, \end{aligned}$$

where

$$\text{(1.7)}\ \alpha_k = \frac{1}{\delta}, \quad \beta_k = -\frac{1}{\omega_{k+1}\delta}, \quad \gamma_k = \frac{1}{\omega_{k+1}\delta} - \frac{1}{\delta}, \quad k = 0, 1, 2, \ldots .$$

Let $\lambda(A)$ denote the spectrum of $A$. It is easy to show that the iterates $\boldsymbol{x}_k$ converge to the solution $\boldsymbol{x}^* = A^{-1}\boldsymbol{b}$ of 1.1 as $k$ increases if the interval $[a, b]$ contains $\lambda(A)$; see [7,9,11]. Moreover, consider two intervals $[a, b]$ and $[a', b']$, and assume that $\lambda(A) \subset [a, b] \subset [a', b']$. Then the asymptotic rate of convergence for the interval $[a, b]$ is faster than for the interval $[a', b']$. Therefore, in order to make Chebyshev iteration for the solution of 1.1 converge as quickly as possible, good estimates of the smallest and largest eigenvalues of $A$ are required. Chebyshev iteration is often implemented as an adaptive algorithm, in which a suitable interval is determined during the iterations; see [9] for a discussion. Recently, Golub and Kent [6] described a new adaptive Chebyshev iteration method. In this method inner products of residual vectors are interpreted as modified moments. The modified moments and recursion coefficients 1.7 yield a symmetric tridiagonal matrix $\tilde{T}_m$, whose extreme eigenvalues are used to determine the interval $[a, b]$ for Chebyshev iteration. We review this scheme in Section 2.

The eigenvalues of $\tilde{T}_m$ are the nodes of a Gaussian quadrature rule. The present paper describes how the weights of this quadrature rule can be used also. We propose that the eigenvalues of smallest magnitude of $\tilde{T}_m$ should be ignored when updating the interval $[a, b]$ if the associated weights are "tiny". This strategy for updating the interval is based on the observation that the Gaussian quadrature rule provides an approximation of the spectral density function defined by $A$ and $\boldsymbol{r}_0$. Details of our scheme are presented in Section 3. A different but related application of Gaussian quadrature rules is described by Fischer and Freund [3], who consider the determination of polynomial preconditioners.

A question of interest is how sensitive the nodes of a Gaussian quadrature rule are to perturbations in the modified moments. These perturbations may be caused, e.g., by a perturbation in the residual vector $\boldsymbol{r}_0$. We show in Section 4 that nodes associated with small weights are sensitive to perturbations. This provides another motivation for ignoring Gaussian nodes associated with small weights. Computed examples are presented in Section 5. These examples compare our new adaptive Chebyshev iteration method with the adaptive scheme described in [6] and show that our modification can yield significantly fewer iterations.

Sometimes it is convenient to use a splitting $A = M - N$, such that both $M$ and $N$ are symmetric, $M$ is positive definite and the linear systems of equations $M\boldsymbol{a} = \boldsymbol{c}$, for $\boldsymbol{c} \in \mathrm{R}^n$, can be solved rapidly. We then solve

$$(I - M^{-1}N)\boldsymbol{x} = M^{-1}\boldsymbol{b} \tag{1.1$'$}$$

instead of 1.1. Modified moments can also be defined for iterative methods applied to the solution of 1.1′; see Golub and Kent [6] for details.

## 2. Application of modified moments to Chebyshev iteration.

This section describes the adaptive Chebyshev iteration scheme due to Golub and Kent [6] and introduces notation necessary to present our new adaptive scheme. Introduce the spectral resolution

$$A = Q\Lambda Q^T, \tag{2.1}$$

where

$$\begin{aligned} \Lambda &= \mathrm{diag}[\lambda_1, \lambda_2, \ldots, \lambda_n], \qquad \lambda_1 \le \lambda_2 \le \cdots \le \lambda_n, \\ Q &= [\boldsymbol{q}_1, \boldsymbol{q}_2, \ldots, \boldsymbol{q}_n], \qquad Q^T Q = I. \end{aligned}$$

Let $\boldsymbol{x}_0$ be an initial approximate solution of 1.1 and write the residual vector $\boldsymbol{r}_0 = \boldsymbol{b} - A\boldsymbol{x}_0$ as a linear combination of eigenvectors

$$\frac{\boldsymbol{r}_0}{\|\boldsymbol{r}_0\|} = \sum_{j=1}^{n} \gamma_j \boldsymbol{q}_j, \quad \gamma_j = \boldsymbol{q}_j^T \boldsymbol{r}_0 / \|\boldsymbol{r}_0\|, \tag{2.2}$$

where $\|\cdot\|$ denotes the Euclidean vector norm. Introduce the spectral density function

$$d\sigma_n(\lambda)=\sum_{j=1}^{n}\delta(\lambda-\lambda_j)\gamma_j^2\,, \tag{2.3}$$

where $\lambda_j$ and $\gamma_j$ are defined by 2.1–2.2 and $\delta(\lambda)$ is the Dirac $\delta$-function. We refer to $d\sigma_n$ as the spectral density function of $A$ associated with $\boldsymbol{r}_0$. The cumulative distribution function $\sigma_n$ associated with $d\sigma_n$ is piecewise constant with a jump $\gamma_j^2$ at $\lambda_j$.

Introduce the inner product

$$(f,g)=\int_R f(\lambda)g(\lambda)d\sigma_n(\lambda)=\sum_{j=1}^{n}f(\lambda_j)g(\lambda_j)\gamma_j^2. \tag{2.4}$$

Following Golub and Kent [6], we introduce modified moments associated with the Chebyshev iteration method. From 2.2, 2.4, 1.3 and the orthogonality of the vectors $\boldsymbol{q}_j$, we obtain

$$\frac{\boldsymbol{r}_k^T\boldsymbol{r}_l}{\|\boldsymbol{r}_0\|^2}=\sum_{j=1}^{n}p_k(\lambda_j)p_l(\lambda_j)\gamma_j^2=\int_R p_k(\lambda)p_l(\lambda)d\sigma_n(\lambda)=(p_k,p_l)\,.$$

In particular, we can define the modified moments

$$\nu_k=\frac{\boldsymbol{r}_k^T\boldsymbol{r}_0}{\|\boldsymbol{r}_0\|^2}=\int_R p_k(\lambda)d\sigma_n(\lambda)=(p_k,1)\,. \tag{2.5}$$

It follows from 2.5 that one modified moment can be computed in each iteration. Golub and Kent [6] point out that, in fact, two modified moments can be computed in each iteration 1.2 by using the following identities for the residual polynomials

$$p_{2k}(\lambda) = p_k^2(\lambda)+\frac{1}{C_{2k}\left(\frac{1}{\mu}\right)}\left[p_k^2(\lambda)-1\right], \tag{2.6}$$

$$p_{2k+1}(\lambda) = p_k(\lambda)p_{k+1}(\lambda)+\frac{1}{\mu C_{2k}\left(\frac{1}{\mu}\right)}\left[p_k(\lambda)p_{k+1}(\lambda)-p_1(\lambda)\right], \tag{2.7}$$

which follow from trigonometric identities. Equations 2.6–2.7 yield

$$\nu_{2k} = (\boldsymbol{r}_k,\boldsymbol{r}_k)+\frac{1}{C_{2k}\left(\frac{1}{\mu}\right)}\left[(\boldsymbol{r}_k,\boldsymbol{r}_k)-\nu_0\right], \tag{2.8}$$

$$\nu_{2k+1} = (\boldsymbol{r}_k,\boldsymbol{r}_{k+1})+\frac{1}{\mu C_{2k}\left(\frac{1}{\mu}\right)}\left[(\boldsymbol{r}_k,\boldsymbol{r}_{k+1})-\nu_1\right]. \tag{2.9}$$

Formulas 2.8 and 2.9 can be used to determine two modified moments in each iteration.

Let $\{\hat{p}_j\}_{j=0}^n$ be a family of orthogonal polynomials with respect to the measure $d\sigma_n$. The modified Chebyshev algorithm determines the recursion coefficients $\hat{\alpha}_k$, $\hat{\beta}_k$ and $\hat{\gamma}_k$ in the three-term recursion formula

$$\begin{aligned} \lambda\hat{p}_k(\lambda) &= \hat{\beta}_k\hat{p}_{k+1}(\lambda) + \hat{\alpha}_k\hat{p}_k(\lambda) + \hat{\gamma}_k\hat{p}_{k-1}(\lambda), \qquad k = 0, 1, 2, \dots, \\ \hat{p}_0(\lambda) &= 1, \quad \hat{p}_{-1}(\lambda) = 0 \end{aligned}$$

From the modified moments $\nu_k$ and the recursion coefficients $\alpha_k$, $\beta_k$ and $\gamma_k$ for the residual polynomials $p_k$; see 1.6. The modified Chebyshev algorithms is described, e.g., in [4,6].

ALGORITHM 2.1 (MODIFIED CHEBYSHEV ALGORITHM).

**Input**: $m$, $\{\nu_k\}_{k=0}^{2m-1}$, $\{\alpha_k\}_{k=0}^{2m-1}$, $\{\beta_k\}_{k=0}^{2m-1}$, $\{\gamma_k\}_{k=0}^{2m-1}$;
**Output**: $\{\hat{\alpha}_k\}_{k=0}^{m-1}$, $\{\hat{\beta}_k\}_{k=0}^{m-1}$, $\{\hat{\gamma}_k\}_{k=0}^{m-1}$;
**for** $j := 0, 1, \dots, 2m-1$ **do**
  $\sigma_{-1,j} := 0$; $\sigma_{0,j} := \nu_j$;
**end** $j$;
$\hat{\alpha}_0 := \alpha_0 + \beta_0 \frac{\sigma_{01}}{\sigma_{00}}$; $\hat{\beta}_0 := \beta_0$; $\hat{\gamma}_0 := 0$;
**for** $k := 1, 2, \dots, m-1$ **do**
  **for** $j := k, k+1, \dots, 2m-k-1$ **do**
    $\sigma_{kj} := [\beta_j\sigma_{k-1,j+1} + (\alpha_j - \hat{\alpha}_{k-1})\sigma_{k-1,j} - \gamma_j\sigma_{k-1,j-1} - \hat{\gamma}_{k-1}\sigma_{k-2,j}]/\hat{\beta}_{k-1}$;
  **end** $j$;
  $\hat{\alpha}_k := \alpha_k + \beta_k \frac{\sigma_{k,k+1}}{\sigma_{kk}} - \beta_{k-1}\frac{\sigma_{k-1,k}}{\sigma_{k-1,k-1}}$; $\hat{\beta}_k := \beta_k$; $\hat{\gamma}_k := \beta_{k-1}\frac{\sigma_{kk}}{\sigma_{k-1,k-1}}$;
**end** $k$; □

We assume that $m$ in Algorithm 2.1 is chosen small enough so that $\sigma_{kk} \neq 0$ for $0 \leq k < m$. The recursion coefficients $\hat{\alpha}_k$, $\hat{\beta}_k$ and $\hat{\gamma}_k$ computed by the algorithm determine the tridiagonal matrix

$$\hat{T}_m = \begin{bmatrix} \hat{\alpha}_0 & \hat{\beta}_0 & & & & \mathbf{0} \\ \hat{\gamma}_1 & \hat{\alpha}_1 & \hat{\beta}_1 & & & \\ & \hat{\gamma}_2 & \hat{\alpha}_2 & \hat{\beta}_2 & & \\ & & & & \ddots & \\ & & & \ddots & \ddots & \hat{\beta}_{m-2} \\ \mathbf{0} & & & & \hat{\gamma}_{m-1} & \hat{\alpha}_{m-1} \end{bmatrix},$$

whose eigenvalues $\{\tilde{\lambda}_j\}_{j=1}^m$ are the zeros of $\hat{p}_m$. For notational simplicity, we assume throughout this paper that

$$0 < \tilde{\lambda}_1 \leq \tilde{\lambda}_2 \leq \dots \leq \tilde{\lambda}_m. \tag{2.10}$$

Let

$$\hat{D}_m = \operatorname{diag}\left[1, \left(\frac{\hat{\gamma}_1}{\hat{\beta}_0}\right)^{1/2}, \left(\frac{\hat{\gamma}_1\hat{\gamma}_2}{\hat{\beta}_0\hat{\beta}_1}\right)^{1/2}, \ldots, \left(\prod_{k=1}^{m-1}\frac{\hat{\gamma}_k}{\hat{\beta}_{k-1}}\right)^{1/2}\right].$$

Then the matrix

$$\tilde{T}_m = \hat{D}_m^{-1}\hat{T}_m\hat{D}_m$$

is symmetric and its eigenvalues agree with those of $\hat{T}_m$. Golub [5] shows that

$$\min\{\lambda : \lambda \in \lambda(A)\} \leq \tilde{\lambda}_1, \qquad \tilde{\lambda}_m \leq \max\{\lambda : \lambda \in \lambda(A)\};$$

see also Szegő [10, Chapter 3] for related results. These inequalities led Golub and Kent [6] to propose to update the interval $[a, b]$ now and then in the following manner. Let $[a', b']$ denote the updated interval. Its end points are defined by

$$a' := \min\{a, \tilde{\lambda}_1\}, \tag{2.11}$$

$$b' := \max\{b, \tilde{\lambda}_m\}. \tag{2.12}$$

After each update Chebyshev iteration is restarted using Chebyshev polynomials for the updated interval. Computed examples with different strategies for when to update are presented in [6] and in Section 5.

The eigenvalues of $\tilde{T}_m$ are nodes of a Gaussian quadrature rule. Our new updating scheme, described in Section 3, differs from the scheme by Golub and Kent [6] in that the updating not only depends on the nodes of the Gaussian quadrature rule associated with $\tilde{T}_m$ but also on the weights of this quadrature rule. Specifically, the end point $a$ of the interval is reduced only if the Gaussian weight associated with an eigenvalue of $\tilde{T}_m$ smaller than $a$ is sufficiently large.

**3. A new adaptive Chebyshev iteration scheme.** The eigenvalues $\tilde{\lambda}_j$ of $\tilde{T}_m$ are the nodes and the square of the first component of the eigenvectors of unit length yield the weights $\tilde{w}_j$ of an $m$-point Gaussian quadrature rule associated with the density function 2.3,

$$G_m f = \sum_{j=1}^{m} f(\tilde{\lambda}_j)\tilde{w}_j. \tag{3.1}$$

The nodes and weights of the quadrature rule 3.1 can conveniently be computed from $\tilde{T}_m$ by the Golub–Welsch [8] algorithm. We note that if $m = n$ then $\tilde{\lambda}_j = \lambda_j$ and $\tilde{w}_j = \gamma_j^2$ for $1 \leq j \leq n$. The density function

$$d\sigma_m(\lambda) = \sum_{j=1}^{m} \delta(\lambda - \tilde{\lambda}_j)\tilde{w}_j \tag{3.2}$$

associated with 3.1 provides an approximation of $d\sigma_n$ in the following sense:

(i) the first $2m$ moments associated with $d\sigma_m$ and $d\sigma_n$ are the same;
(ii) between each pair of adjacent Gaussian nodes $\{\tilde{\lambda}_j, \tilde{\lambda}_{j+1}\}$ there lies an eigenvalue of $A$, see [5];
(iii) the distribution function $\sigma_n$ grows rapidly between three adjacent nodes if $\sigma_m$ does, where $\sigma_m$ denotes the cumulative distribution function associated with $d\sigma_m$.

Property (iii) is stated in a precise manner by the Separation Theorem shown independently by Chebyshev, Markoff and Stieltjes.

LEMMA 3.1 (SEPARATION THEOREM [10, P. 50]). *Let the nodes $\tilde{\lambda}_j$ be arranged in increasing order. Then there are numbers $\eta_j \in \mathrm{R}$, $a < \eta_1 < \eta_2 < \ldots < \eta_{m-1} < b$, such that*

$$\tilde{w}_j = \sigma_n(\eta_j) - \sigma_n(\eta_{j-1}), \qquad 1 \leq j \leq m,$$

*and*

$$\tilde{\lambda}_j < \eta_j < \tilde{\lambda}_{j+1}, \qquad 1 \leq j < m,$$

*where $\eta_0 = \lambda_1$ and $\eta_m = \lambda_n$.*

Three proofs and further details can be found in [10]. In order to determine the $\eta_j$ of the lemma it may be necessary to modify $\sigma_n$ at some of its jump discontinuities.

It follows from Lemma 3.1 that $\tilde{w}_j \leq \sigma_n(\tilde{\lambda}_{j+1}) - \sigma_n(\tilde{\lambda}_{j-1})$, and, therefore, a large weight $\tilde{w}_j$ implies that $\sigma_n$ grows rapidly in the interval $I_j = \left[\tilde{\lambda}_{j-1}, \tilde{\lambda}_{j+1}\right]$. The growth of $\sigma_n$ may depend on that $I_j$ contains many eigenvalues of $A$, or that $\boldsymbol{r}_0$ contains large components of eigenvectors associated with eigenvalues in $I_j$.

Our scheme for the updating $[a, b]$ depends on both the nodes $\tilde{\lambda}_j$ and the weights $\tilde{w}_j$; the left end point is reduced only if the weight associated with a node $\tilde{\lambda}_j$ smaller than $a$ is sufficiently large. We denote the interval obtained after updating by $[a', b']$. First consider the case $\tilde{\lambda}_1 \geq a$. Then, similarly with 2.11, we let $a' := a$. We turn to the case when $\tilde{\lambda}_1 < a \leq \tilde{\lambda}_2$. It follows from 1.4–1.5 that

$$\max_{a \leq \lambda \leq b} |p_k(\lambda)| = |C_k(1/\mu)|^{-1}. \tag{3.3}$$

Moreover, for $0 < \tilde{\lambda}_1 < a$, we have

$$\max_{a \leq \lambda \leq b} |p_k(\lambda)| < p_k(\tilde{\lambda}_1) < 1. \tag{3.4}$$

Assume for the moment that the Gaussian nodes $\{\tilde{\lambda}_j\}_{j=0}^m$ are eigenvalues of $A$ and that $\{\tilde{\boldsymbol{q}}_j\}_{j=1}^m$ are the associated orthonormal eigenvectors, i.e., $A\tilde{\boldsymbol{q}}_j = \tilde{\boldsymbol{q}}_j\tilde{\lambda}_j$. Let $\boldsymbol{r}_0$ be a linear combination of the $\tilde{\boldsymbol{q}}_j$, i.e.,

$$\frac{\boldsymbol{r}_0}{||\boldsymbol{r}_0||} = \sum_{j=1}^{m} \tilde{w}_j^{1/2} \tilde{\boldsymbol{q}}_j. \tag{3.5}$$

Then

$$\boldsymbol{r}_k = p_k(A)\boldsymbol{r}_0 = \|\boldsymbol{r}_0\| \sum_{j=1}^{m} \tilde{w}_j^{1/2} p_k(\tilde{\lambda}_j)\tilde{\boldsymbol{q}}_j.$$

By 3.4, the component $\tilde{w}_1^{1/2}\tilde{\boldsymbol{q}}_1$ of $\boldsymbol{r}_0$ is damped less after k iterations than the components $\tilde{w}_j^{1/2}\tilde{\boldsymbol{q}}_j$, $2 \le j \le m$. Indeed, in order to achieve an asymptotically optimal rate of convergence by the Chebyshev iteration scheme, the interval $[a,b]$ has to contain all the $\tilde{\lambda}_j$. On the other hand, the components $\tilde{w}_j^{1/2}\tilde{\boldsymbol{q}}_j$, $2 \le j \le m$, are damped faster if the left end point $a$ is not reduced. The representation 3.5 suggests the following strategy for choosing $a'$. Let $k$ be the number of iterations to be carried out before the interval $[a',b']$ is to be updated again. If $\tilde{w}_1$ is sufficiently small so that

$$\tilde{w}_1^{1/2} p_k(\tilde{\lambda}_1) \le \max_{2\le j\le m} \tilde{w}_j^{1/2} \max_{a\le\lambda\le b} |p_k(\lambda)|,$$

then $a' := a$; otherwise let $a' := \tilde{\lambda}_1$. This approach of determining $a'$ seeks to keep the updated interval small until the component of $\boldsymbol{r}_k$ associated with $\tilde{\boldsymbol{q}}_1$ is as large as the other components.

The scheme described for determining $a'$ concerns the case $\tilde{\lambda}_1 \le a < \tilde{\lambda}_2$. If instead $\tilde{\lambda}_j < a \le \tilde{\lambda}_{j+1}$ for an arbitrary $j$, then we similarly as above seek to determine whether the lower bound of the interval should be $\tilde{\lambda}_l$ for $l = j, j-1, \ldots, 1$. The decision depends on the size of the weights associated with these nodes. Algorithm 3.1 below is for this general case.

In many computed examples the weight associated with $\tilde{\lambda}_m$ was found to be large, often $\tilde{w}_m \ge 1/4$. Therefore, the approach used to determine $a'$ would, when applied to the computation of $b'$, in general, reduce to 2.12. This is our motivation for updating the end point $b$ according to 2.12.

ALGORITHM 3.1 (UPDATING THE INTERVAL [A,B]).

***Input***: $a, b, k, \{\tilde{\lambda}_j\}_{j=1}^m, \{\tilde{w}_j^{1/2}\}_{j=1}^m$; *(Assume* $\tilde{\lambda}_1 \le \tilde{\lambda}_2 \le \ldots \le \tilde{\lambda}_m$ *)*
***Output***: $a', b'$;
$b' := \max\{b, \tilde{\lambda}_m\}$; $a' := a$;
***for*** $j := m-1, m-2, \ldots, 1$ ***if*** $\tilde{\lambda}_j < a'$ ***do***
  $\delta := \frac{2}{b'+a'}$; $\mu := \frac{b'-a'}{b'+a'}$; $\lambda_j' := 1 - \delta\tilde{\lambda}_j$;
  ***if*** $\tilde{w}_j^{1/2} C_k(\frac{\lambda_j'}{\mu}) > \max_{j+1\le l\le m} \{\tilde{w}_l^{1/2}\}$ ***then*** $a' := \tilde{\lambda}_j$;
***end*** $j$; □

The inequality in Algorithm 4.1 follows from 3.3 and 3.4. The evaluation of $C_k(\frac{\lambda_j'}{\mu})$ is carried out by 1.5.

**4. A sensitivity analysis.** This section is concerned with the sensitivity of the nodes $\tilde{\lambda}_j$ in the Gaussian quadrature rule 3.1 to perturbations in the modified moments $v_k$, given by 2.5. Our analysis is closely related to an analysis by Gautschi [4] of the condition number of the mapping $F$

defined below. We have

$$\sum_{j=1}^{m} p_k(\tilde{\lambda}_j)\tilde{w}_j = \nu_k, \qquad 0 \le k < 2m.$$

Introduce the map $F$: $\mathrm{R}^{2m} \to \mathrm{R}^{2m}$; $\{\tilde{w}_1, \ldots, \tilde{w}_m, \tilde{\lambda}_1, \ldots, \tilde{\lambda}_m\} \to \{v_k\}_{k=0}^{2m-1}$. Its Jacobian is given by $J = PW$, where

$$P = \begin{bmatrix} p_0(\tilde{\lambda}_1) & \cdots & p_0(\tilde{\lambda}_m) & p_0'(\tilde{\lambda}_1) & \cdots & p_0'(\tilde{\lambda}_m) \\ p_1(\tilde{\lambda}_1) & \cdots & p_1(\tilde{\lambda}_m) & p_1'(\tilde{\lambda}_1) & \cdots & p_1'(\tilde{\lambda}_m) \\ \vdots & & \vdots & \vdots & & \vdots \\ p_{2m-1}(\tilde{\lambda}_1) & \cdots & p_{2m-1}(\tilde{\lambda}_m) & p_{2m-1}'(\tilde{\lambda}_1) & \cdots & p_{2m-1}'(\tilde{\lambda}_m) \end{bmatrix}$$

and

$$W = \mathrm{diag}[1, 1, \ldots, 1, \tilde{w}_1, \tilde{w}_2, \ldots, \tilde{w}_m].$$

The Jacobian of $F^{-1}$ is given by $J^{-1} = W^{-1}P^{-1}$. Let $dy$ denote the differential of $y$. Then

$$\begin{bmatrix} d\tilde{w}_1 \\ \vdots \\ d\tilde{w}_m \\ d\tilde{\lambda}_1 \\ \vdots \\ d\tilde{\lambda}_m \end{bmatrix} = W^{-1}P^{-1} \begin{bmatrix} d\nu_1 \\ d\nu_2 \\ \vdots \\ d\nu_m \end{bmatrix}. \tag{4.1}$$

From 4.1 it follows that $d\tilde{\lambda}_j$ is proportional to $1/\tilde{w}_j$. This suggests that nodes $\tilde{\lambda}_j$ associated with "tiny" weights $\tilde{w}_j$ are sensitive to perturbations in the modified moments. These nodes may therefore be particularly sensitive to changes in the initial vector $\boldsymbol{x}_0$.

**5. Computed examples.** In this section we discuss the results of some numerical experiments in which the performance of the adaptive Chebyshev iteration method described in this paper is compared with the performance of the scheme proposed by Golub and Kent in [6]. All programs are written in FORTRAN 77. The numerical experiments described were carried out on an IBM RISC 6000/550 workstation using double precision arithmetic, i.e., with approximately 15 significant digits. We derive a sequence of test problems by discretizing the Poisson equation

$$\Delta u = f \tag{5.1}$$

on the unit square $\Omega := \{(x, y) : 0 \leq x, y \leq 1\}$, and impose boundary condition $u(x, y) = 0$ on $\partial\Omega$. The right hand side function $f$ is chosen so that is $u(x, y) = xe^{xy}\sin(\pi x)\sin(\pi y)$ solves the boundary value problem. We discretize $\Omega$ by a uniform $(n+2) \times (n+2)$ grid, including boundary points, and approximate $\Delta$ by the standard five-point stencil. This generates a symmetric linear system of $N := n^2$ equations in $n^2$ unknowns $u_{ij}$, $1 \leq i, j \leq n$. The quantity $u_{ij}$ approximates the solution $u$ of the boundary value problem at the grid point $(ih, jh)$, where $h = \frac{1}{n+1}$. We scale the linear system obtained in this manner by $h^2$ and write it as 1.1.

We remark that unless an accurate initial approximate solution $\boldsymbol{x}_0$ is available the linear systems of equations considered can be solved more rapidly by a fast Poisson solver than by our iterative scheme. The advantage of our scheme is that it can be applied when fast Poisson solvers cannot. The main purpose of the computed examples is to compare our scheme with the method in [6]. No preconditioner is used, in order to keep the issues of interest clear. However, in practical applications it is often desirable to use a preconditioner.

The adaptive Chebyshev algorithm based on the the use of Gaussian weights is denoted by "WCHB" in the tables. The adaptive Chebyshev scheme by Golub and Kent [6] is referred to as "CHB". We recall that the two schemes only differ in the way the left end point of the interval $[a, b]$ is updated. The column labeled "maxupdt" shows how many times approximations of eigenvalues of $A$ are computed, and the column labeled "frequency" shows how often they are determined. For example, if maxupdt=20 and frequency=15, then approximate eigenvalues are calculated every 15 iterations but at most 20 times. The column labeled "changes in a" displays how many times the value of the left end point changes during the iterations. This number is, in general, equal to maxupdt in CHB. After maxupdt updates of the interval, the left end point $a$ is set to the least of all computed eigenvalue estimates. This value of $a$ and the current value of $b$, which is the largest of all computed eigenvalue estimates, define an interval $[a, b]$. We apply nonadaptive Chebyshev iteration with respect to this interval until the residual error is sufficiently small. In the computed examples the iterations are terminated when the residual error is of norm less than $1 \cdot 10^{-5}$.

In all examples described the number $m$ of eigenvalues calculated in each adaptive step is chosen to be 5. Thus, $2m = 10$ modified moments are computed during the first 5 Chebyshev iterations after each update of the interval $[a, b]$, using the formulae 2.8–2.9. The Gaussian quadrature rule associated with the tridiagonal matrix $\tilde{T}_5$ generated by the modified Chebyshev algorithm is used to update the interval in the scheme labeled WCHB. In the CHB scheme only the extreme eigenvalues of $\tilde{T}_5$ are applied. We occasionally experienced numerical stability problems in the modified Chebyshev algorithm when $m$ was chosen larger than 5.

In order to start the Chebyshev iteration method an initial interval $[a, b]$

TABLE 5.1
$N = 6,400$

| adaptive method | maxupdt | frequency | updates of a | steps |
|---|---|---|---|---|
| WCHB | 20 | 15 | 2 | 767 |
| WCHB | 25 | 15 | 6 | 766 |
| WCHB | 30 | 15 | 11 | 820 |

is required. We assume that a suitable interval for Chebyshev iteration is not known before beginning the iterations. In all examples described $\frac{1}{n}\text{trace}(A) = 4$ and we chose $[3.9, 4.1]$ as our initial interval. A good choice of initial interval $[a, b]$ is such that $\lambda_1 \leq a < b \leq \lambda_n$, because the adaptive algorithm only decreases $a$ and increases $b$ during the iterations. If for the initial interval $[a, b]$, we have $\lambda_1 \ll a < b \ll \lambda_n$, then it may be desirable to update $a$ and $b$ as soon as possible, in order to avoid having the norm of the residual vector become too large. In our computed examples we therefore updated the initial interval already after $m = 5$ iterations, independently of the parameter frequency. The initial vector $\boldsymbol{x}_0$ is a random vector with elements uniformly distributed in the interval $[-1, 1]$. We remark that numerical experiments suggest that the performance of both algorithms CHB and WCHB is not very sensitive to the choices of initial interval and of $\boldsymbol{x}_0$.

The implementations of the adaptive schemes WCHB and CHB contain many elements necessary in a black-box adaptive Chebyshev iteration method. Our computed examples therefore shed light on how such a black-box scheme should be designed. In fact, our implementations WCHB and CHB would qualify as black-box solvers if equipped with an adaptive strategy for how to determine the parameter maxupdt. Such a strategy would have to be based on a comparison of the predicted decrease of the residual error, assuming that all eigenvalues lie in the interval determined by the smallest and largest computed eigenvalue estimate, and the actual decrease in the norm of the residual error. Our computational experience indicates that quite a few updates of the interval $[a, b]$ should be carried out before the interval is kept fixed. Typically a fairly large number of iterations are required to carry out these updates. Our scheme WCHB can reduce the number of these iterations significantly, compared with CHB.

Example 5.1. We solve the Poisson equation 5.1 on a uniform grid with $80 \times 80$ interior grid points, i.e, the order of $A$ is $n = 6400$. Tables 5.1–5.2 show the number of Chebyshev iteration steps required to reduce the Euclidean norm of the residual vector to less than $10^{-5}$ as a function of the number of times we estimate eigenvalues of $A$. In the tables the number of iterations for both schemes WCHB and CHB is smallest for maxupdt=25. Numerous computed examples suggest that frequency=15

TABLE 5.2
$N = 6,400$

| adaptive method | maxupdt | frequency | steps |
|---|---|---|---|
| CHB | 20 | 15 | 915 |
| CHB | 25 | 15 | 906 |
| CHB | 30 | 15 | 953 |

TABLE 5.3
$N = 10,000$

| adaptive method | maxupdt | frequency | updates of a | steps |
|---|---|---|---|---|
| WCHB | 15 | 15 | 2 | 985 |
| WCHB | 20 | 15 | 2 | 902 |
| WCHB | 25 | 15 | 4 | 907 |
| WCHB | 30 | 15 | 9 | 960 |

is a good choice. For these parameter values CHB requires approximately 18% more iterations than WCHB.

If maxupdt is reduced to 20, then the determined interval yields slower convergence than when maxupdt=25. This depends on that the interval $[a, b]$ determined after 20 updates is smaller than the interval obtained after 25 updates. On the other hand, maxupdt=30 requires more iterations than maxupdt=25 because the additional restarts do not improve the interval, but make it necessary to use Chebyshev polynomials of lower degree. □

Example 5.2. In this example we solve Poisson's equation 5.1 on a uniform grid with $100 \times 100$ interior grid points, i.e., $n = 10,000$. In Tables 5.3–5.4 the fastest convergence of WCHB and CHB is achieved for maxupdt= 20. The smallest number of iterations with CHB is about 17% larger than the smallest number of iterations with WCHB. □

Example 5.3. We solve the linear system 1.1 obtained by discretizing the Poisson equation 5.1 on a uniform grid with $120 \times 120$ interior grid points. We remark that, as the order of $A$ increases, the left end point $a$ of the determined interval moves closer to the origin. The smallest number of iterations displayed in Tables 5.5–5.6 is for maxupdt=30. □

**6. Conclusion.** Chebyshev iteration is based on the use of scaled Chebyshev polynomials for an interval on the positive real axis. This paper describes a new adaptive scheme for determining a suitable interval. Similarly as Golub and Kent [6], we use modified moments to determine a symmetric tridiagonal matrix. In [6] the eigenvalues of the tridiagonal ma-

TABLE 5.4
$N = 10,000$

| adaptive method | maxupdt | frequency | steps |
|---|---|---|---|
| CHB | 15 | 15 | 1104 |
| CHB | 20 | 15 | 1056 |
| CHB | 25 | 15 | 1076 |
| CHB | 30 | 15 | 1124 |

TABLE 5.5
$N = 14,400$

| adaptive method | maxupdt | frequency | updates of a | steps |
|---|---|---|---|---|
| WCHB | 25 | 15 | 3 | 1160 |
| WCHB | 30 | 15 | 3 | 1151 |
| WCHB | 35 | 15 | 3 | 1181 |

TABLE 5.6
$N = 14,400$

| adaptive method | maxupdt | frequency | steps |
|---|---|---|---|
| CHB | 25 | 15 | 1414 |
| CHB | 30 | 15 | 1338 |
| CHB | 35 | 15 | 1393 |

trix are used to determine the interval; we apply the Gaussian quadrature rule associated with the tridiagonal matrix. Computed examples illustrate that this modification can reduce the number of iterations required considerably and therefore should be considered when developing black-box adaptive Chebyshev iterative solvers.

**Acknowledgement.** Work by D.C. and L.R. on the present paper was carried out during a visit to IPS at ETH. They would like to thank Martin Gutknecht for making this visit possible.

## REFERENCES

[1] D. Calvetti, G.H. Golub and L. Reichel, Adaptive Chebyshev iterative methods for nonsymmetric linear systems based modified moments, Report ICM-9205-30, Institute for Computational Mathematics, Kent State University, Kent, OH, May 1992.

[2] J.J. Dongarra, I.S. Duff, D.C. Sorensen and H.A. van der Vorst, *Solving Linear Systems on Vector and Shared Memory Computers*, SIAM, Philadelphia, PA, 1991.

[3] B. Fischer and R.W. Freund, On adaptive weighted polynomial preconditioning for Hermitian positive definite matrices, Report 92.09, RIACS, NASA Ames Research Center, Moffett Field, CA.

[4] W. Gautschi, On generating orthogonal polynomials, *SIAM J. Sci. Stat. Comput.*, **3** (1982), pp. 289–317.

[5] G.H. Golub, Some modified matrix eigenvalue problems, *SIAM Review*, **15** (1974), pp. 318–334.

[6] G.H. Golub and M.D. Kent, Estimates of eigenvalues for iterative methods, *Math. Comp.*, **53** (1989), pp. 619–626.

[7] G.H. Golub and R.S. Varga, Chebyshev semi-iterative methods, successive overrelaxation methods, and second order Richardson iterative methods, *Numer. Math.*, **3** (1961), pp. 147–168.

[8] G.H. Golub and J.H. Welsch, Calculation of Gauss quadrature rules, *Math. Comp.*, **23** (1969), pp. 221–230.

[9] L.A. Hageman and D.M. Young, *Applied Iterative Methods*, Academic Press, New York, NY, 1981.

[10] G. Szegő, *Orthogonal Polynomials*, 4th ed., Amer. Math. Soc., Providence, RI, 1975.

[11] R.S. Varga, *Matrix Iterative Analysis*, Prentice Hall, Englewood Cliffs, NJ, 1962.

# ORDERING EFFECTS ON RELAXATION METHODS APPLIED TO THE DISCRETE CONVECTION-DIFFUSION EQUATION

HOWARD C. ELMAN * AND MICHAEL P. CHERNESKY †

**Abstract.** We present an analysis of relaxation methods for the discrete convection-diffusion equation based on norms of the iteration matrices. For one-dimensional problems, the results show how the performance of iterative solvers is affected by directions of flow associated with the underlying operator. In particular, for problems of size $n$, relaxation sweeps opposite the direction of flow incur a latency of approximately $n$ steps in which convergence is slow, and red-black relaxation incurs a latency of approximately $n/2$ steps. There is no latency associated with relaxation that follows the flow. The one-dimensional analysis is also generalized to two-dimensional problems in the case where relaxation follows the flow.

**1. Introduction.** Consider the linear system of equations

$$Au = b \tag{1.1}$$

where $A$ is a nonsingular matrix. Let $A = D - L - U$, where $D$ is a diagonal or block-diagonal matrix, and $L$ and $U$ are the strict lower-triangular and upper-triangular parts of $D - A$, respectively. The Gauss-Seidel iterative method for solving (1.1) is

$$u^{(k+1)} = (D-L)^{-1}Uu^{(k)} + (D-L)^{-1}b,$$

where $u^{(0)}$ is an initial guess. The traditional convergence analysis of this method is based on the spectral radius of the iteration matrix, $\rho((D-L)^{-1}U)$; the iteration is convergent if and only if $\rho((D-L)^{-1}N) < 1$, and roughly speaking, for large $k$ the error decreases in magnitude by a factor of $\rho((D-L)^{-1}U)$ at each step [10],[11].

We are interested in linear systems (1.1) that arise from discretization of the constant-coefficient convection-diffusion equation with Dirichlet boundary conditions,

$$-\Delta u + v \cdot \nabla u = f \text{ on } \Omega, \qquad u = g \text{ on } \partial\Omega, \tag{1.2}$$

on some domain $\Omega \subset R^d$, $d = 1$ or 2. In this case, precise definitions of the matrices $D$, $L$, and $U$ depend on an ordering of the underlying grid. Possible ordering strategies include "natural" and "red-black" orderings [11]. Moreover, for each of these, there are several variants that depend

* Department of Computer Science and Institute for Advanced Computer Studies, University of Maryland, College Park, MD 20742, e-mail: elman@cs.umd.edu. This work was supported by the U. S. Army Research Office under grant DAAL-0392-G-0016, and by the National Science Foundation under grants ASC-8958544 and CCR-8818340.

† Department of Mathematics, University of Maryland, College Park, MD 20742, email: chernesk@cs.umd.edu.

on the orientation of the underlying grid with respect to the direction of flow, determined by the first order term of (1.2). For example, for one-dimensional problems on an interval with (the scalar) $v > 0$, if the natural left-to-right ordering of grid points is used, then the sweeps associated with Gauss-Seidel iteration "follow" the flow, whereas with the right-to-left ordering, the sweeps "go against" the flow.

Among natural orderings, it is generally believed that choices for which the relaxation sweeps follow the flow are superior; see e.g. [2],[4],[5],[6],[9]. Relaxation with red-black orderings tends to be somewhat less effective than sweeping with the flow but more effective than sweeping against the flow [4],[5]. The traditional analysis of relaxation methods for consistently ordered problems makes use of the relation $\rho((D-L)^{-1}U) = \rho(D^{-1}(L+U))^2$ between the Gauss-Seidel and Jacobi iteration matrices. However, $\rho(D^{-1}(L+U))$ is independent of the ordering and signs of $v$. so that, although the analysis reveals the asymptotic behavior of such methods as the number of iterations grows, it provides no information about their transient behavior.

In this paper, we present an analysis of relaxation methods for the convection-diffusion equation that captures the differences among ordering strategies. For one-dimensional problems, we outline bounds on the matrix $l_1$-norm $\|((D-L)^{-1}U)^k\|_1$, and we show that these bounds largely predict the effects of flow direction and orderings, of the Gauss-Seidel method. These results are a summary of material presented in [3]. In addition, we generalize the analysis to two-dimensional problems in the case of relaxation sweeps that follow the flow. These results show, for several discretization strategies, that sweeping with the flow also incurs no latency in the two-dimensional case.

An outline of the rest of the paper is as follows. In §2, we present a set of numerical examples and convergence results that explains the behavior of the Gauss-Seidel method for different orderings, for the one-dimensional problem. In §3, we present the analysis and experimental results for the two-dimensional case, and in §4, we discuss generalizations of the analysis to other norms.

**2. One-dimensional problems.** Consider the one-dimensional version of (1.2) on the unit interval $(0,1)$, where $v$ is now a scalar, $v = \sigma$. Without loss of generality, we will assume that $\sigma > 0$. Most discretizations on a uniform grid of width $h = \frac{1}{n+1}$ lead to a linear system (1.1) in which, for a natural ordering, the coefficient matrix $A$ is tridiagonal of order $n$, with constant values on its three interior bands. We will assume that $A$ is normalized to have unit diagonal, so that it can be represented as

$$A = tri\,[-b,\, 1,\, -c]\,.$$

In addition, we require $b + c = 1$ and $b > 0$, $c > 0$. The first condition is necessary for the discretization to be consistent, and the two inequalities ensure that the discrete solution is not oscillatory [7]. For example, for centered finite difference or piecewise linear finite element discretizations, $b = \frac{1}{2}\left(1 + \frac{\sigma h}{2}\right)$, $c = \frac{1}{2}\left(1 - \frac{\sigma h}{2}\right)$; and for the upwind finite difference discretization, $b = (1+\sigma h)/(2+\sigma h)$, $c = 1/(2+\sigma h)$. We say that the discrete problem is *convection-dominated* if $b$ is large, i.e., close to 1. The Gauss-Seidel iteration matrix is $\mathcal{L}_1 = (I - L)^{-1}U$, where $L$ and $U$ are the strict lower triangular and upper triangular parts of $A$.

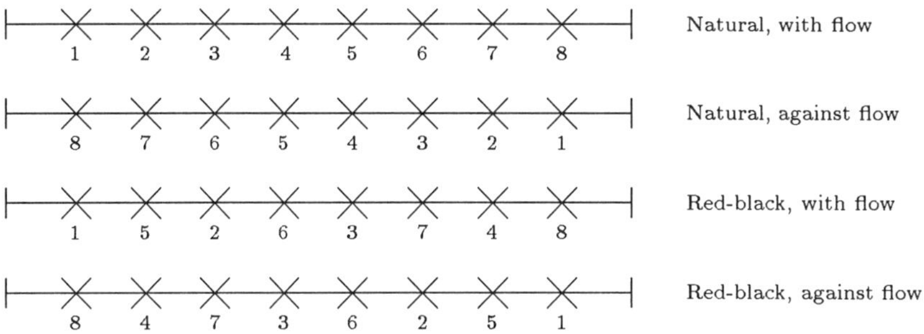


FIG. 2.1. *Four orderings for a one-dimensional grid, for* $\sigma > 0$ *and* $n = 8$.

Let $n$ denote the number of interior mesh points, and let $e^{(k)}$ denote the error $u-u^{(k)}$ at the $k$th step of relaxation. Figure 2.1 shows examples of four different orderings for $n = 8$. There are two natural orderings, together with two red-black orderings induced by the natural orderings. Figure 2.2 shows a representative example of the behavior of relaxation for convection-dominated problems that reveals the limitations of the classical analysis. The figure plots $||e^{(k)}||_1$, on a logarithmic scale, against the iteration count $k$, for the Gauss-Seidel method corresponding to the four ordering schemes. Here, $n = 32$ and $b = 7/8$. The initial guess is a normally distributed random vector with mean 0 and variance 1, and the right hand side and solution are identically zero. The spectral radius for each of the orderings is $\rho(\mathcal{L}_1) = .434$. Figure 2.3 shows the norms $||\mathcal{L}_1^k||_1$. In both figures, the highlighted values correspond to $k = n-1$ and $k = n/2-1$ for the natural ordering against the flow and red-black orderings, respectively. It is evident that the norms are closely correlated with the performance of the solution algorithm, and that the spectral radius reveals nothing about the transient behavior.

REMARK 1. The particular problem used for Fig. 2.2 corresponds to various choices of $\sigma$ in (1.2), depending on the discretization. For example, for centered differences, it corresponds to $\sigma = 49.5$, and cell Reynolds number $\sigma h/2 = 3/4$. For upwind differences, it corresponds to to $\sigma = 198$ and $\sigma h = 6$.

The iteration matrices arising from different orderings will be distinguished as follows. For the left-to-right natural ordering, inducing a relax-

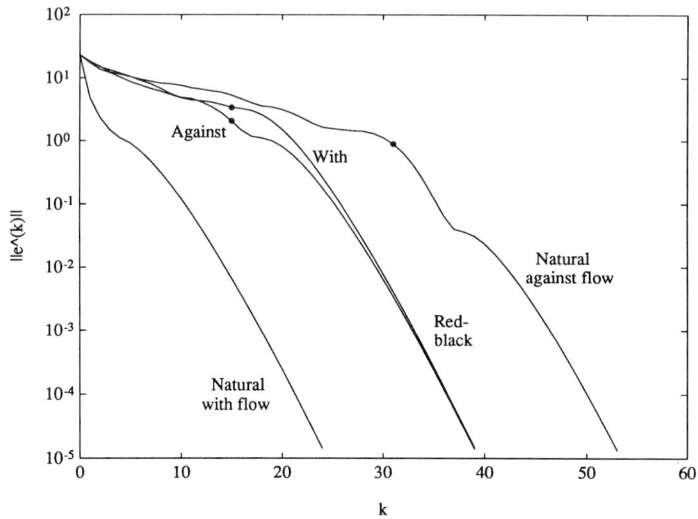


FIG. 2.2. *$l_1$-norms of the errors in Gauss-Seidel iteration, for $n = 32$ and $b = 7/8$.*

ation sweep oriented with the flow, the iteration matrix is $F = (I - L)^{-1}U$, where

$$L = tri\,[b,\, 0,\, 0], \quad U = tri\,[0,\, 0,\, c].$$

The red-black ordering induced by this natural ordering gives rise to the coefficient matrix $A = I - L_{RB} - U_{RB}$ where

$$L_{RB} = \begin{pmatrix} 0 & 0 \\ B & 0 \end{pmatrix}, \quad U_{RB} = \begin{pmatrix} 0 & C \\ 0 & 0 \end{pmatrix},$$

and

$$B = tri\,[0,\, b,\, c], \quad C = tri\,[b,\, c,\, 0],$$

of dimensions $\lceil n/2 \rceil \times \lfloor n/2 \rfloor$ and $\lfloor n/2 \rfloor \times \lceil n/2 \rceil$ respectively. The iteration matrix is given by

$$F_{RB} = (I - L_{RB})^{-1}U_{RB} = (I + L_{RB})U_{RB} = \begin{pmatrix} 0 & C \\ 0 & BC \end{pmatrix}.$$

For sweeps oriented against the flow, rather than reversing the ordering, it is equivalent to use the left-to-right natural ordering and perform an "upper-triangular" sweep, i.e., with the iteration matrix $G = (D - U)^{-1}L$.

We summarize the convergence analysis for the one-dimensional problem as follows. There are three results: *lower bounds* on the values of both $||G^k||_1$ and $||F^k_{RB}||_1$, and *upper bounds* on the values of $||F^k||_1$.

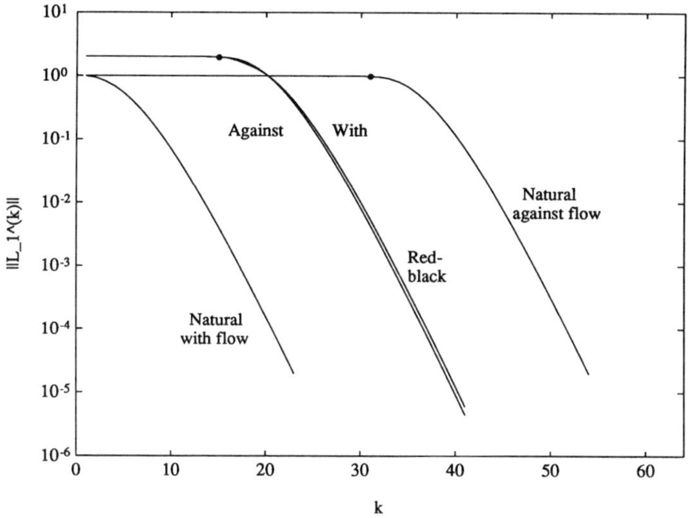


FIG. 2.3. $\log_{10} \|\mathcal{L}_1^k\|_1$ *for* $n = 32$ *and* $b = 7/8$.

THEOREM 2.1. *The norm* $\|G^k\|_1$ *for Gauss-Seidel iteration with sweeps against the flow is bounded below for* $k < n$ *by*

$$\|G^k\|_1 \geq (1-c^2)^{k-1}(1-c^{n-(k-1)}).$$

THEOREM 2.2. *For problems whose order* $n$ *is divisible by four, the norm* $\|F_{RB}^k\|_1$ *for Gauss-Seidel iteration associated with the red-black ordering induced by a left-to-right natural ordering is bounded below as follows:*

$$\|F_{RB}^k\|_1 \geq \begin{cases} 2 & \text{for } k < n/4, \\ 2 - c^{n/2} & \text{for } k = n/4, \\ \alpha_k \phi_0(k,c) + \beta_k \phi_1(k,c) + \gamma_k \phi_2(k,c) & \text{for } n/4 < k \leq n/2 - 1, \end{cases}$$

*where*

$$\phi_0(k,c) = (1-c)^{2(k-n/4)}(2-c^{n/2}),$$
$$\phi_1(k,c) = c(1-c)^{2(k-n/4)-1}(1-c^{n/2-2}) + c(1-c)^{2(k-n/4)}(1-c^{n/2-1}),$$
$$\phi_2(k,c) = c^2(1-c)^{2(k-n/4)-1}(1-c^{n/2-3}) + c^2(1-c)^{2(k-n/4)}(1-c^{n/2-2}),$$

*and*

$$\alpha_{n/4+j} = 1, \ \beta_{n/4+j} = 2j, \ \gamma_{n/4+j} = \begin{cases} j(2j-1) & \text{if } j < n/4 - 1 \\ j(2j-1) - 1 & \text{if } j = n/4 - 1 \end{cases}.$$

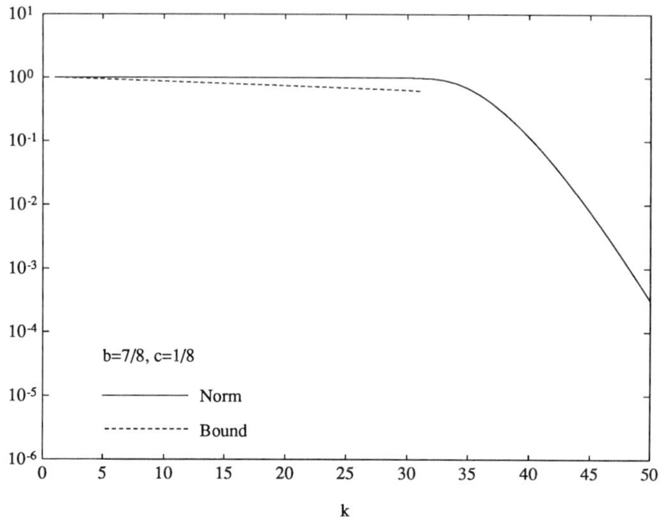


FIG. 2.4. *Comparison of* $\|G^k\|_1$ *with lower bounds, for* $n = 32$ *and* $c = 1/8$.

THEOREM 2.3. *The norm* $\|F^k\|_1$ *for Gauss-Seidel iteration with sweeps that follow the flow is bounded above by*

$$\|F^k\|_1 \leq 1 - b^{n+k-2}\left(\sum_{j=0}^{k-1}\binom{n+k-3+j}{j}c^j\right). \tag{2.1}$$

*For* $k > (n-3)c/(1-2c)$, *the following simpler upper bound holds:*

$$\|F^k\|_1 \leq \frac{k\binom{n+2k-3}{k}b^{n-2}(bc)^k}{k(1-2c)-(n-3)c}. \tag{2.2}$$

REMARK 2. The proof of this theorem depends on the fact that $F$ has the form

$$F = c\begin{pmatrix} 0 & 1 & & & \\ 0 & b & \ddots & & \\ \vdots & \vdots & \ddots & 1 & \\ 0 & b^{n-2} & \cdots & b & 1 \\ 0 & b^{n-1} & \cdots & b^2 & b \end{pmatrix};$$

see [3].

The results of Theorems 2.1 and 2.2 indicate that if $c$ is near one, then the norms of the iteration matrices for sweeping against the flow and the

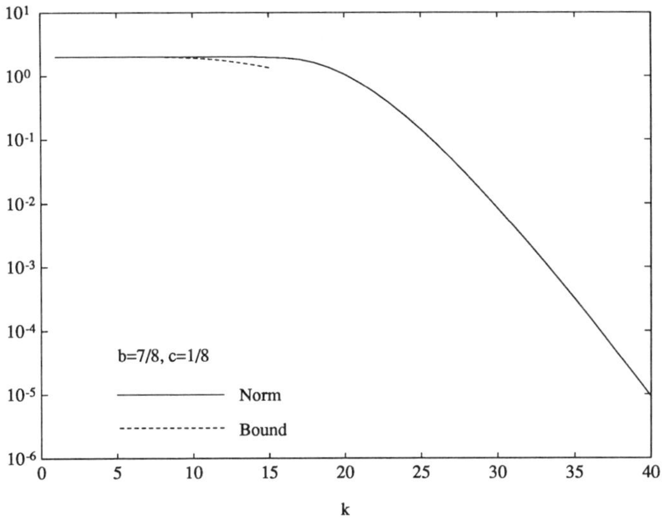


FIG. 2.5. *Comparison of* $||F_{RB}^k||_1$ *with lower bounds, for* $n = 32$ *and* $c = 1/8$.

red-black ordering are close to one for $n-1$ and $n/2-1$ steps, respectively. Consequently, these orderings incur a latency in which the little reduction in the error is obtained. In contrast, Theorem 2.3 indicates that the norm of the iteration matrix is small for sweeping with the flow. Figures 2.4, 2.5 and 2.6 plots the norms of the iteration matrices and the bounds, for $n = 32$, $b = 7/8$, and $c = 1/8$. (This is the same problem considered in Figs. 2.2 and 2.3.) Proofs of these results and descriptions of additional numerical experiments are given in [3].

**3. Two-dimensional problems.** In this section, we show that the analysis of sweeping with the flow for one-dimensional problems [3], summarized in Theorem 2.3, can be generalized to the two-dimensional case where $v = (\sigma, \tau)$ is a constant vector. We will assume that $\Omega = (0,1) \times (0,1)$, and that the discretization is made on a uniform grid of width $h = \frac{1}{n+1}$ such that for a natural ordering of the grid, the coefficient matrix has block-tridiagonal form

$$A = tri\,[-L_l,\, D_l,\, -U_l\,]\,. \tag{3.1}$$

That is, $A = D - L - U$ where $D$ is a block diagonal matrix each of whose diagonal blocks is equal to $D_l$, $L$ is a lower triangular matrix in which each entry of the first subdiagonal block band is $-L_l$, and $U$ is a similarly defined upper triangular matrix. The problem (1.2) has associated with it a direction of flow given by $d_f = (\sigma, \tau)$. We will also assume that the ordering producing (3.1) corresponds to a line ordering of the underlying grid, so that the induced Gauss-Seidel iteration also has a sweep direction $d_s$ associated with it. For example, Fig. 3.1 shows a horizontal line ordering of an $8 \times 8$ grid, with lines ordered from bottom to top; here, $d_s = (0, 1)$.

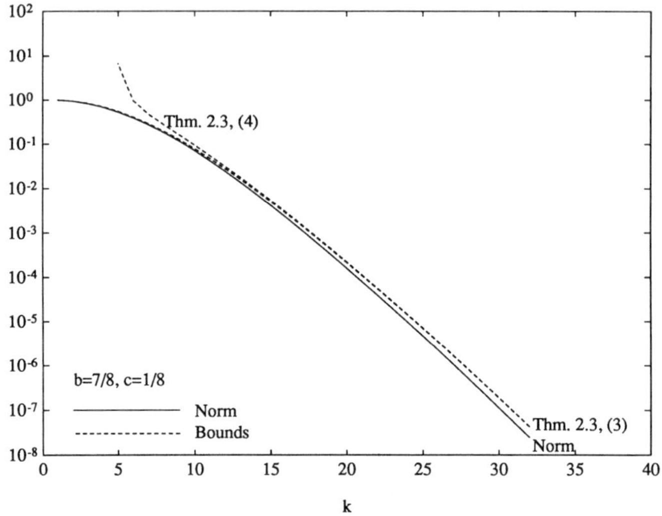


FIG. 2.6. *Comparison of* $||F^k||_1$ *with upper bounds, for* $n = 32$ *and* $b = 7/8$.

Let $\theta$ denote the angle formed by the vectors $d_f$ and $d_s$, $0° \le \theta \le 180°$. If $0° \le \theta < 90°$, we say that the Gauss-Seidel iteration sweeps are "with the flow," if $90° < \theta \le 180°$, then they are "against the flow," and if $\theta = 90°$, then they are orthogonal to the direction of flow.

The analysis is based on the following result.

LEMMA 3.1. *Let* $M$ *denote a matrix grouped into blocks* $M_{ij}$. *If* $||M_{ij}||_1 \le m_{ij}$ *for each* $(i,j)$, *then* $||M||_1 \le ||\widehat{M}||_1$, *where* $\widehat{M} = [m_{ij}]$.

The iteration matrix for the Gauss-Seidel sweep associated with a natural ordering is

$$F = \begin{pmatrix} 0 & (D_l^{-1}U_l) & & & \\ 0 & (D_l^{-1}L_l)(D_l^{-1}U_l) & \ddots & & \\ \vdots & \vdots & \ddots & (D_l^{-1}U_l) & \\ 0 & (D_l^{-1}L_l)^{n-2}(D_l^{-1}U_l) & \cdots & (D_l^{-1}L_l)(D_l^{-1}U_l) & (D_l^{-1}U_l) \\ 0 & (D_l^{-1}L_l)^{n-1}(D_l^{-1}U_l) & \cdots & (D_l^{-1}L_l)^2(D_l^{-1}U_l) & (D_l^{-1}L_l)(D_l^{-1}U_l) \end{pmatrix}.$$

The nonzero blocks of $F$ satisfy

$$||(D_l^{-1}L_l)^k(D_l^{-1}U_l)||_1 \le \hat{b}^k\hat{c},$$

where $\hat{b} = ||D^{-1}||_1||L||_1$ and $\hat{c} = ||D^{-1}||_1\,||U||_1$. Thus, by Lemma 3.1,

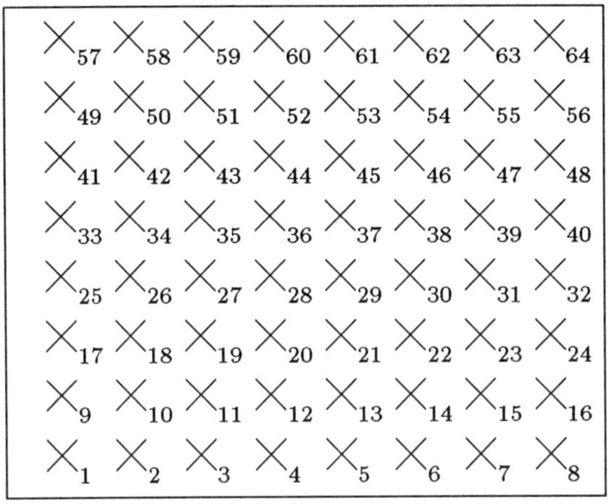


FIG. 3.1. *Horizontal line ordering for a two-dimensional* $8 \times 8$ *grid.*

$||F^k||_1 \le ||\widehat{F}^k||_1$, where

$$\widehat{F} = \hat{c} \begin{pmatrix} 0 & 1 & & & \\ 0 & \hat{b} & \ddots & & \\ \vdots & \vdots & \ddots & 1 & \\ 0 & \hat{b}^{n-2} & \dots & \hat{b} & 1 \\ 0 & \hat{b}^{n-1} & \dots & \hat{b}^2 & \hat{b} \end{pmatrix}.$$

In light of Remark 2 of §2, we have the following result.

THEOREM 3.2. *If* $||F^k||_1 \le ||\widehat{F}^k||_1$ *as in Lemma 3.1, and*

$$\hat{b} \ge 0, \ \hat{c} \ge 0, \text{ and } \hat{b} + \hat{c} = 1, \tag{3.2}$$

*then for Gauss-Seidel iteration applied to systems with coefficient matrix of the form (3.1), the bounds (2.1) and (2.2) apply, with* $\hat{b}$ *and* $\hat{c}$ *in place of* $b$ *and* $c$*, respectively.*

We will give examples below of several discretizatons that satisfy (3.2).

Suppose the discretization of (1.2) is such that the matrices on the block diagonal of $A$ are tridiagonal of the form

$$D_l = tri\,[-c,\, a,\, -d]\,, \tag{3.3}$$

where

$$c \ge 0, \quad d \ge 0, \text{ and } a \ge c + d, \tag{3.4}$$

i.e., $D_l$ is a diagonally dominant $M$-matrix [10, p. 85]. Then, $D_l$ satisfies

$$D_l = \hat{D}_l + E,$$

where

$$\hat{D}_l = \hat{L}_l \hat{\Delta}_l \hat{U}_l = \begin{pmatrix} 1 & & & & \\ -\frac{c}{\delta} & 1 & & & \\ & & \ddots & & \\ & & -\frac{c}{\delta} & 1 & \\ & & & -\frac{c}{\delta} & 1 \end{pmatrix} \begin{pmatrix} \delta & & & & \\ & \delta & & & \\ & & \ddots & & \\ & & & \delta & \\ & & & & \delta \end{pmatrix} \begin{pmatrix} 1 & -\frac{d}{\delta} & & & \\ & 1 & -\frac{d}{\delta} & & \\ & & \ddots & & \\ & & & 1 & -\frac{d}{\delta} \\ & & & & 1 \end{pmatrix},$$

and $\delta = \frac{1}{2}(a + \sqrt{a^2 - 4cd})$. $E$ is identically zero except for the $(1,1)$-entry

$$a - \delta = \frac{1}{2}(a - \sqrt{a^2 - 4cd}).$$

But (3.4) implies that $0 \le cd \le \frac{1}{4}a^2$, so that $E$ is real and nonnegative. Therefore, the $M$-matrices $D_l$ and $\hat{D}_l$ satisfy $D_l \ge \hat{D}_l$, which implies $0 < D_l^{-1} \le \hat{D}_l^{-1}$ [10, p. 87]. Consequently,

$$\|D_l^{-1}\|_1 \le \|\hat{D}_l^{-1}\|_1 \le \|\hat{U}_l^{-1}\|_1 \|\hat{\Delta}_l^{-1}\|_1 \|\hat{L}_l^{-1}\|_1 < \frac{1}{a-(c+d)}. \tag{3.5}$$

As an example, consider a five-point finite difference operator, where $A$ has the form (3.1) with $D_l$ as in (3.3), $L_l = bI$, $U_l = eI$, and, in addition to (3.4), the inequalities

$$b \ge 0, \quad e \ge 0, \text{ and } a = b + c + d + e.$$

This is true for centered finite differences provided that the cell Reynolds numbers $\sigma h/2$ and $\tau h/2$ are bounded in absolute value by 1, and for upwind finite differences. It follows from (3.5) that

$$\hat{b} = \frac{b}{a-(c+d)}, \qquad \hat{c} = \frac{e}{a-(c+d)},$$

so that $\hat{b} + \hat{c} = 1$ and Theorem 3.2 can be applied. For example, with centered finite differences, we have

$$\hat{b} = \frac{1}{2}\left(1 + \frac{\tau h}{2}\right), \quad \hat{c} = \frac{1}{2}\left(1 - \frac{\tau h}{2}\right). \tag{3.6}$$

Fig. 3.2 shows two examples of the bounds of Theorem 3.2, for a $16 \times 16$ grid and $\hat{b} = 7/8$, $\hat{c} = 1/8$. This corresponds to the centered difference discretization with $\tau h/2 = 3/4$. The first graph shows the case $\sigma h/2 = 0$. In this case, flow is in the vertical ($y$) direction, and the relaxation sweeps are parallel to the direction of flow ($\theta = 0$). The second graph shows the case $\sigma h/2 = 1/2$; the angle between the flow direction and sweep direction is approximately 34°. As in §2, the bounds are in general agreement with the actual norms, although when the sweeps are not parallel to the flow

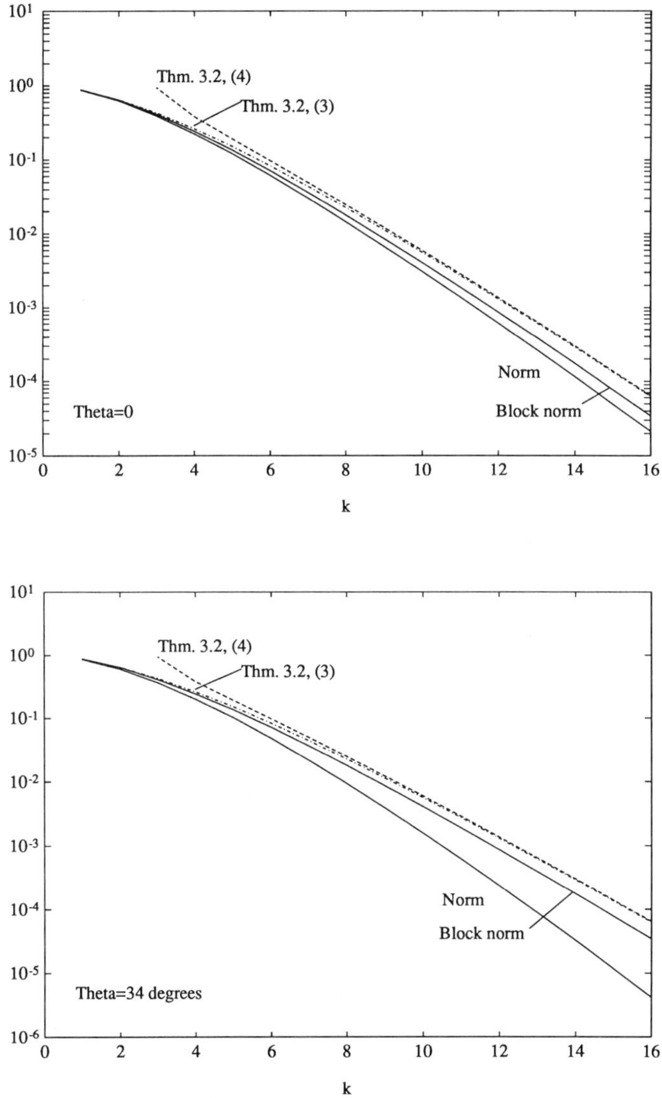


FIG. 3.2. *Comparison of* $||F^k||_1$ *with upper bounds, for centered finite differences,* $n = 16$ *and* $\hat{b} = 7/8$. *Top:* $\theta = 0$, *bottom:* $\theta \approx 34°$.

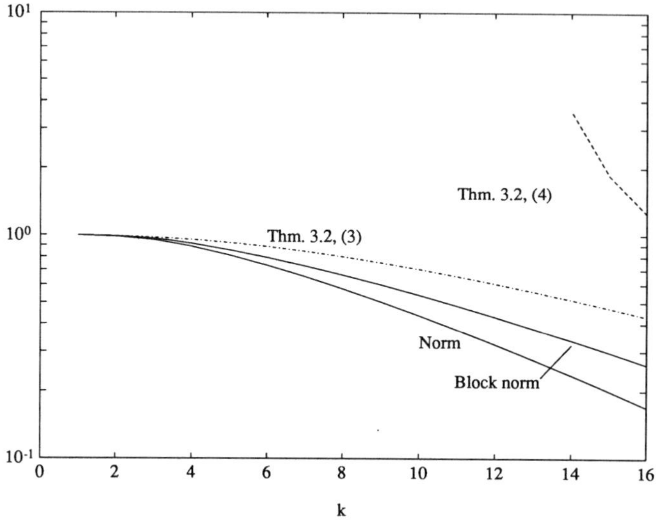


FIG. 3.3. *Comparison of* $||F^k||_1$ *with upper bounds, for bilinear finite elements,* $n = 16$, $\hat{b} = 2/3$, $\theta = 0$.

(as in the second graph), there is some degradation in the quality of the analysis caused by the use of the block norm.

For a second example, let $A$ come from a finite element discretization with bilinear elements on squares [8]. In this case,

$$\begin{aligned} D_l &= tri\left[-(1+\sigma h),\, 8,\, -(1-\sigma h)\right], \\ L_l &= tri\left[-(1+\tfrac{(\sigma+\tau)h)}{4}),\, -(1+\tau h),\, -(1-\tfrac{(\sigma-\tau)h)}{4})\right], \\ U_l &= tri\left[-(1+\tfrac{(\sigma-\tau)h)}{4}),\, -(1-\tau h),\, -(1-\tfrac{(\sigma+\tau)h)}{4})\right]. \end{aligned}$$

If all the off-diagonal entries of $A$ are nonpositive, then

$$||D_l^{-1}||_1 \le \frac{1}{6}, \quad ||L_l||_1 = 3 + \frac{3}{2}\tau h, \quad ||U_l||_1 = 3 - \frac{3}{2}\tau h.$$

This is the case provided $|\sigma h| \le 1$ and $|\tau h| \le 1$. Theorem 3.2 is satisfied with $\hat{b}$ and $\hat{c}$ exactly as in (3.6). Note that the requirements on $\sigma h$ and $\tau h$ are more restrictive than for centered finite differences, and in particular, the restriction on $\tau h$ limits the maximum value of $\hat{b}$ to $3/4$. Figure 3.3 shows an example of the bounds for $n = 16$, $\sigma h = 0$ and $\tau h = 2/3$, giving $\hat{b} = 2/3$, $\hat{c} = 1/3$ and $\theta = 0$. This example is less dramatically convection-dominated than the previous example, and the norms of the iteration matrix decrease less rapidly than when $\hat{b}$ is closer to 1. Here, the first bound from Theorem 3.2 qualitatively agrees with the computed norms, but the second bound of the analysis does not provide useful information.

**4. Additional remarks.** The lower bounds of Theorems 2.1 and 2.2 also apply if the $l_\infty$-norm is used in place of the $l_1$-norm [3]. We have also observed empirically for two-dimensional problems that for discretizations on an $n \times n$ grid, block relaxation with a natural ordering against the flow incurs a latency of approximately $n$ steps, and block red-black relaxation incurs a latency of approximately $n/2$ steps. We know of no analytic results that predict this.

The upper bounds of Theorems 2.3 and 3.2 also apply for the $l_\infty$-norm. As a consequence, using [1, p. 13]

$$||F||_p = ||F||_1^{1/p} ||F||_\infty^{1/q},$$

where $\frac{1}{p} + \frac{1}{q} = 1$, the bounds also apply for any $l_p$-norm, and in particular for $l_2$.

## REFERENCES


[1] J. Bergh and J. Löfström, *Interpolation Spaces, An Introduction*, Springer-Verlag, Berlin, 1976.

[2] M. Eiermann, *Fields of Values and Iterative Methods*, tech. report, IBM Scientific Center, Heidelberg, 1991.

[3] H. C. Elman and M. P. Chernesky, *Ordering effects on relaxation methods applied to the discrete convection-diffusion equation*, Tech. Report UMIACS-TR-92-40, Institute for Advanced Computer Studies, University of Maryland, 1992.

[4] H. C. Elman and G. H. Golub, *Iterative methods for cyclically reduced non-self-adjoint linear systems, II*, Math. Comp., 56 (1991), pp. 215–242.

[5] ———, *Line iterative methods for cyclically reduced convection-diffusion problems*, SIAM J. Sci. Stat. Comput., 13 (1992), pp. 339–363.

[6] P. A. Farrell, *Flow conforming iterative methods for convection dominated flows*, in Numerical and Applied Mathematics, C. Brezinski, ed., J. C. Baltzer AG, Scientific Publishing Co., 1989, pp. 681–686.

[7] P. M. Gresho and R. L. Lee, *Don't suppress the wiggles – they're telling you something*, Computers and Fluids, 9 (1981), pp. 223–253.

[8] C. Johnson, *Numerical Solution of Partial Differential Equations by the Finite Element Method*, Cambridge University Press, New York, 1987.

[9] L. N. Trefethen, *Pseudospectra of Matrices*, Tech. Report 91-10, Oxford University Computing Laboratory, 1991.

[10] R. S. Varga, *Matrix Iterative Analysis*, Prentice-Hall, Englewood Cliffs, New Jersey, 1962.

[11] D. M. Young, *Iterative Solution of Large Linear Systems*, Academic Press, New York, 1970.

# ON THE ERROR COMPUTATION FOR POLYNOMIAL BASED ITERATION METHODS

BERND FISCHER * AND GENE H. GOLUB †

**Abstract.** In this note we investigate the Chebyshev iteration and the conjugate gradient method applied to the system of linear equations $Ax = f$ where $A$ is a symmetric, positive definite matrix. For both methods we present algorithms which approximate during the iteration process the $k$th error $\varepsilon_k = \|x - x_k\|_A$ . The algorithms are based on the theory of modified moments and Gaussian quadrature. The proposed schemes are also applicable for other polynomial iteration schemes. Several examples, illustrating the performance of the described methods, are presented.

**1. Introduction.** Consider the system of linear equations

$$Ax = f, \tag{1.1}$$

where $A$ is a $N \times N$ real symmetric, positive definite matrix and $f$ is a given vector. Let $x_0 \in \mathbb{R}^N$ be any initial guess for the solution of (1.1), and let $r_0 := f - Ax_0$, $\varepsilon_0 := x - x_0$ be the associated residual vector and error vector, respectively. Furthermore, let

$$\mathcal{P}_n := \{q(t) \equiv a_0 + a_1 t + \cdots + a_n t^n \mid a_0, a_1, \ldots, a_n \in \mathbb{R}\}$$

denote the set of all real polynomials of degree at most $n$.

A widely used class of iterative schemes for the solution of (1.1) are the so-called *polynomial iteration methods.* These methods generate iterates of the form

$$x_n = x_0 + q_{n-1}(A) r_0, \quad \text{where} \quad q_{n-1} \in \mathcal{P}_{n-1}, \quad n = 1, 2, \ldots .$$

The corresponding residual vectors and error vectors are given in terms of the so-called residual polynomial $p_n$

$$\begin{aligned} r_n &= f - Ax_n = p_n(A) r_0, \\ \varepsilon_n &= x - x_n = p_n(A) \varepsilon_0, \end{aligned} \quad \text{where} \quad p_n(t) \equiv 1 - t q_{n-1}(t). \tag{1.2}$$

Notice that $\varepsilon_n = A^{-1} r_n$ and consequently

$$\|\varepsilon_n\|_A^2 := \varepsilon_n^T A \varepsilon_n = r_n^T A^{-1} r_n. \tag{1.3}$$

In this note we are concerned with the approximation of $\|\varepsilon_n\|_A$ as the iteration process goes along.

---

* Institute of Applied Mathematics, University of Hamburg, D–2000 Hamburg, F.R.G.

† Department of Computer Science, Stanford University, Stanford, CA 94305, U.S.A.
The work of this author was in part supported by the National Science Foundation under Grant NSF CCR–8821078.

In §2 we show that the computation of $||\varepsilon_n||_A$ is equivalent to the evaluation of a certain integral. This integral will be (approximately) evaluated by means of Gaussian quadrature. The process is closely related to the computation of certain orthogonal polynomials. In §2.1 and §2.2 we show how to approximate $||\varepsilon_n||_A$ during the *Chebyshev iteration* and the *conjugate gradient (CG) method*, respectively. It turns out that the computation of the initial error $||\varepsilon_0||_A$ is special. §2.3 is devoted to this problem. Finally, in §3 we report on some numerical experiments.

**2. Error computation.** In this section we show that the $A$-norm of the error (1.3) can be expressed in terms of a certain Riemann-Stieltjes integral. To this end we expand the starting residual

$$r_0 = \sum_{k=1}^{N} \sigma_k v_k \tag{2.1}$$

into orthonormal eigenvectors $v_k$ of $A$. We denote by $\lambda_k$ the eigenvalues corresponding to $v_k$ and obtain, in view of (1.3) and (2.1),

$$\begin{aligned} ||\varepsilon_n||_A^2 &= (p_n(A)r_0)^T A^{-1} p_n(A) r_0 = \sum_{k=1}^{N} \frac{\sigma_k^2}{\lambda_k} p_n^2(\lambda_k) \\ &= \int p_n^2(t) d\hat{\sigma}(t). \end{aligned} \tag{2.2}$$

The distribution function $\hat{\sigma}(t)$ in the above defined Riemann - Stieltjes integral is given by

$$\hat{\sigma}(t) \equiv \sum_{k=1}^{N} \frac{\sigma_k^2}{\lambda_k} H(t-\lambda_k), \quad \text{where} \quad H(t) \equiv \begin{cases} 1 & \text{for } t \geq 0, \\ 0 & \text{otherwise.} \end{cases} \tag{2.3}$$

The problem of determining the error $||\varepsilon_n||_A$ is equivalent to the evaluation of an integral with distribution function $\hat{\sigma}(t)$. A standard tool for computing such an integral is the Gaussian quadrature . Let us assume (for a moment) that we know the orthonormal [1] polynomials $\hat{\psi}_k$, $k = 1, 2, \ldots, n$, relative to $\hat{\sigma}(t)$

$$\int \hat{\psi}_k(t)\hat{\psi}_m(t) d\hat{\sigma}(t) \begin{cases} = 1 & \text{if } k = m \\ = 0 & \text{if } k \neq m \end{cases}$$

in terms of their three-term recurrence coefficients $\hat{\alpha}_k$ and $\hat{\beta}_k$

$$t\hat{\psi}_k(t) = \hat{\beta}_k \hat{\psi}_{k+1}(t) + \hat{\alpha}_k \hat{\psi}_k(t) + \hat{\beta}_{k-1} \hat{\psi}_{k-1}(t).$$

[1] the assumption of orthonormality is more convenient but not necessary.

We associate with $\hat{\sigma}(t)$ the tridiagonal matrix

$$\hat{J}_n = \begin{pmatrix} \hat{\alpha}_1 & \hat{\beta}_1 & & & \\ \hat{\beta}_1 & \hat{\alpha}_2 & \hat{\beta}_2 & & \\ & \ddots & \ddots & \ddots & \\ & & \ddots & \ddots & \hat{\beta}_{n-1} \\ & & & \hat{\beta}_{n-1} & \hat{\alpha}_n \end{pmatrix} \tag{2.4}$$

Then, it is well known (see, e.g., Wilf [12]) that the Gaussian quadrature applied to (2.2) reads

$$\int p_n^2(t)d\hat{\sigma}(t) = \hat{\nu}_0 \; e_1^T p_n(\hat{J}_n)p_n(\hat{J}_n)e_1,$$

where $e_1 = (1, 0, \dots, 0)^T$ denotes the first unit vector and

$$\hat{\nu}_0 = \int d\hat{\sigma}(t) = \|\varepsilon_0\|_A^2 \tag{2.5}$$

the zero-order moment, which turns out to be the square of the initial error. Hence, the evaluation of $\|\varepsilon_n\|_A$ "reduces" to the computation of the so-called Jacobi matrix $\hat{J}_n$ and the moment $\hat{\nu}_0$.

The computation of the orthogonal polynomials relative to $\hat{\sigma}(t)$ seems to be on the first glance a quite tricky problem. However, as we will see, it is not hard to compute the orthogonal polynomials

$$\psi_{k+1}(x) = (x - \alpha_k)\psi_k(x) - \beta_k\psi_{k-1}(x), \quad k = 0, 1, \dots, \tag{2.6}$$

with respect to the distribution function (compare (2.3))

$$\sigma(t) \equiv \sum_{k=1}^{N} \sigma_k^2 H(t - \lambda_k). \tag{2.7}$$

Again (cf. (2.4)), we associate with $\sigma(t)$ a Jacobi matrix $J_n$. For the CG-method the $\psi_k(t) \equiv p_k(t)$ are just the (explicitly known) residual polynomials (see §2.2). For general polynomial iteration methods the first $n$ polynomials $\psi_k, k = 1, 2, \dots, n$, can be calculated via a *modified moment algorithm* (see, e.g., Sack, Donovan [10] and Wheeler [11]). This algorithm requires as input (only) the $2n + 1$ modified moments

$$\nu_j = \int p_j(t)d\sigma(t), \quad j = 0, 1, \dots 2n. \tag{2.8}$$

These moments, however, are at hand during the iteration process. More precisely, we have by (1.2) and (2.1) (compare Dahlquist, Eisenstat and

Golub [3]),

$$
(2.9)\qquad \begin{aligned} r_j^T r_0 &= r_0^T p_j(A) r_0 = \sum_{k=1}^{N} \sigma_k^2 p_j(\lambda_k) \\ &= \int p_j(t) d\sigma(t) = \nu_j. \end{aligned}
$$

Thus after $2n$ steps of the polynomial iteration method we can calculate $\psi_1, \psi_2, \ldots, \psi_n$. Obviously, it would be advantageous to have the $n$th degree orthogonal polynomial after only $n$ iteration steps. In §2.2 we will see that this is precisely the case for the Chebyshev iteration.

Anyway, after having computed the $\psi_k$, the desired polynomials can be obtained from a "modification algorithm", by noting that, in view of (2.3) and (2.7), essentially $\hat{\sigma}(t) \equiv \sigma(t)/t$ holds. Such algorithms can be found in Gautschi [6] and Fischer and Golub [4]. Here, for the computation of the Jacobi matrix $\hat{J}_n$ (cf. (2.4)) of order $n$, one needs to know beforehand the zero-order moment $\hat{\nu}_0$ (cf. (2.5)) and the Jacobi matrix $J_n$.

How to economically compute $J_n$ and $\hat{\nu}_0$, respectively, will be explained in the next subsections.

**2.1. The Chebyshev iteration.** The Chebyshev iteration method is a so-called parameter dependent method. The scheme needs estimates, $a$ and $b$, for the smallest and largest eigenvalue of $A$, respectively,

$$
(2.10)\qquad \sigma(A) \subset [a, b], \quad \text{where} \quad 0 < a \le \lambda_{min}(A) \le \lambda_{max}(A) \le b.
$$

The actual computation of these bounds will not be discussed in this paper. We refer the reader to Hageman and Young [9], Golub and Kent [7], Calvetti and Reichel [1], and references therein. The residual polynomial (cf. (1.2)) for the Chebyshev iteration

$$
(2.11)\qquad p_n(t) = \frac{T_n\left((a+b-2t)/(a-b)\right)}{T_n\left((a+b)/(a-b)\right)}
$$

is a suitable scaled and translated Chebyshev polynomial of the first kind. As is well-known, the Chebyshev polynomials fulfill the following identities

$$
\begin{aligned} T_{2n-1}(t) &= 2\, T_n(t) T_{n-1}(t) - t, \\ T_{2n}(t) &= 2\, T_n^2(t) - 1. \end{aligned}
$$

This leads together with (2.9) and (2.11) to the modified moments (compare Golub and Kent [7])

$$
\begin{aligned} \nu_{2n-1} &= r_n^T r_{n-1} + \tfrac{c}{T_{2n-1}(c)}(r_n^T r_{n-1} - \nu_1), \\ \nu_{2n} &= r_n^T r_n + \tfrac{1}{T_{2n}(c)}(r_n^T r_n - \nu_0), \end{aligned} \qquad \text{where} \quad c = \frac{a+b}{a-b}.
$$

Hence, after $n$ steps of the Chebyshev iteration, the modified moments $\nu_0, \nu_1, \ldots, \nu_{2n}$ are available. However, the price paid for the computation of the modified moments are 3 inner products per iteration. The same number of inner products is needed for one step of the conjugate gradient method.

In the next section we show (compare Dahlquist, Golub, and Nash [2]) how to compute the error $||\varepsilon_n||_A$ during the conjugate gradient iteration.

**2.2. The conjugate gradient method.** For the conjugate gradient method, there is, in contrast to general polynomial iteration methods, no additional work required for the computation of the orthogonal polynomials $\psi_k$. This is based on the equivalence of the CG method and the Lanczos algorithm and the fact that the Lanczos process directly computes the desired three-term recurrence coefficients (see, e.g., Golub and Van Loan [8]).

From the minimization property of the CG method one readily obtains [2]

$$||\varepsilon_n||_A^2 = ||\varepsilon_0||_A^2 - \nu_0 e_1^T J_n^{-1} e_1. \tag{2.12}$$

In other words, $||\varepsilon_n||_A^2$ is the error obtained by approximating the initial error $||\varepsilon_0||_A^2 = \int d\hat{\sigma}(t)$ by the $n$-point Gaussian quadrature rule associated with the distribution function $\sigma(t)$.

However, the remaining (and main) problem is the computation of the initial error $||\varepsilon_0||_A$. The next section is devoted to this problem.

**2.3. Computing the initial error.** An effective scheme for the evaluation of

$$||\varepsilon_0||_A^2 = \hat{\nu}_0 = \int d\hat{\sigma}(t)$$

is devised in Gautschi [5]. The scheme is based on the observation that $\hat{\nu}_0$ has a (convergent) *continued fraction* expansion in terms of the three-term recurrence coefficients of the orthogonal polynomial $\psi_k$ (cf. (2.6)):

$$\lim_{k\to\infty} C_k = -||\varepsilon_0||_A^2,$$

where

$$\begin{aligned} C_k &= (1+\rho_k)C_{k-1} - \rho_k C_{k-2}, \\ \rho_{k+1} &= \tfrac{1}{1-\beta_k(1+\rho_k)/(\alpha_k\alpha_{k-1})} - 1, \end{aligned} \qquad k = 1, 2, \ldots . \tag{2.13}$$

As it is not surprising, the convergence rate of (2.13) depends on the condition number of $A$, which determines the (relative) distance between the singular point zero of (2.5) and the interval of integration.

A different approach for the approximation of the initial errror is provided by applying a Gauß-Radau rule onto (2.5) (for details see [2]). This leads to upper and lower bounds for $||\varepsilon_n||_A$ which can be used in conjunction with (2.12). As for the Chebyshev iteration (cf. (2.10)) this scheme requires the knowledge of upper and lower bounds for the spectrum of $A$.

**3. Examples.** In this section we present some numerical examples. The test problems are obtained by discretizing the Poisson equation

$$\Delta u = f$$

on the unit square $\Omega = \{(x,y)|0 \le x, y \le 0\}$ with Dirichlet boundary conditions $u(x,y) = 0$ on $\partial\Omega$. We approximate $\Delta u$ by the standard five-point stencil on a uniform $n \times n$ grid. This results in linear system $Ax = f$ of order $N := n^2$, where the right hand side $f$ is chosen such that the vector $x = (1,1,\ldots,1)^T$ solves the system. In all experiments the initial vector $x_0$ is a random vector with elements uniformly distributed in $[-1,1]$. All computations were carried out in MATLAB.

A crucial point for the correct error estimation is the convergence of the continued fraction (cf. (2.13)) to the initial error $||\varepsilon_0||_A$. In the following table we list the number $K$ of steps needed for the algorithm (2.13) to converge up to machine precision EPS ($\sim 2.22 * 10^{-16}$), i.e.,

$$\frac{|C_K - C_{K-1}|}{|C_K|} < \text{EPS}. \tag{3.1}$$

In order to set the quantity $K$ into perspective we computed in addition the number $L$ of iteration steps required to reduce the (relative) $A$-norm of the error to less than $10^{-5}$, i.e.,

$$\frac{||\varepsilon_L||_A}{||\varepsilon_0||_A} \le 10^{-5}. \tag{3.2}$$

In all examples described we chose for the Chebyshev iteration method the best possible interval $[a = \lambda_{min}(A), b = \lambda_{max}(A)]$ (cf. (2.10)), in order to keep the issues of interest clear. Cond denotes the condition number of the respective systems.

**Table 3.1.** Steps needed for the continued fraction algorithm (2.13) to converge to the initial error $||\varepsilon_0||_A$

| | | CG | | Cheb | |
|---|---|---|---|---|---|
| N | Cond | K | L | K | L |
| 400 | 178 | 62 | 62 | 64 | 123 |
| 900 | 388 | 91 | 89 | 89 | 187 |
| 2500 | 1053 | 141 | 145 | 142 | 309 |

The convergence of algorithm (2.13) is quite rapidly. The next figure shows the typical convergence behaviour. Here, we ploted subsequent approximations $C_k$, $k = 1, 2, \ldots, 10$ (little circles) and the initial error $||\varepsilon_0||_A$ (dotted line) for the CG method and $N = 2500$.

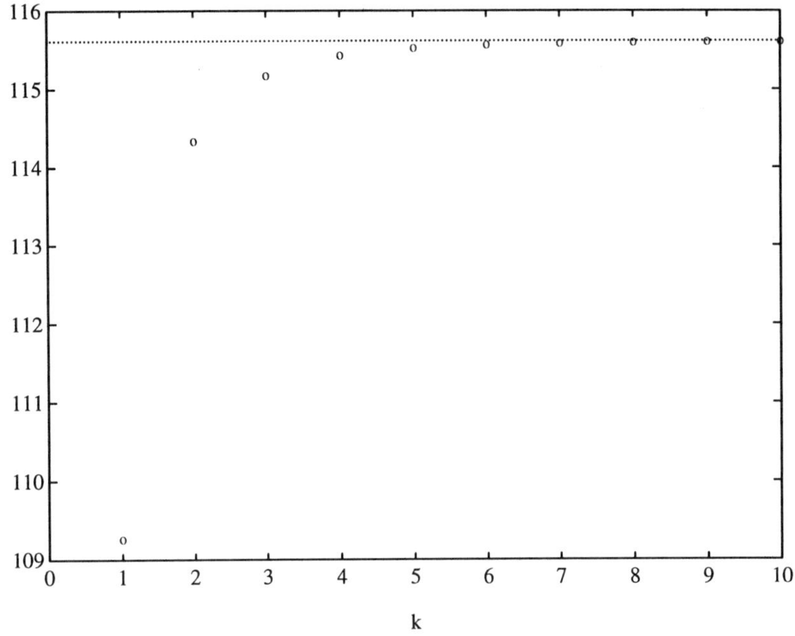


FIG. 3.1. *Approximation of* $||\varepsilon_0||_A$ *by the continued fractions* $C_k$

Furthermore, it should be mentioned that in all our experiences the algorithm (2.13) turned out to be very stable.

In the next table we compare the estimated error $||\varepsilon_k^{est}||_A$ with the true error $||\varepsilon_k||_A$ for the various examples and methods. We list the largest observed error (compare (3.2))

$$\max_{k=1,2,\ldots,L} \left| \frac{||\varepsilon_k^{est}||_A - ||\varepsilon_k||_A}{||\varepsilon_0||_A} \right| .$$

In order to demonstrate the performance of the proposed schemes we first computed the initial error $||\varepsilon_0||_A$ up to machine precision (cf. (3.1)) and then used this value for the subsequent computations.

**Table 3.2.** Maximal deviation of the estimated error from the true error

| N | CG | Cheb |
|---|---|---|
| 400 | $1.21 * 10^{-8}$ | $1.04 * 10^{-8}$ |
| 900 | $1.20 * 10^{-8}$ | $1.20 * 10^{-6}$ |
| 2500 | $1.15 * 10^{-8}$ | $2.47 * 10^{-5}$ |

It is apparent that the error estimates for the Chebyshev iteration become worse with increasing order. This is due to instabilities in the modified moment algorithm. Note, that the computations for the case $N = 2500$ involve orthogonal polynomials of degree 309. However, in practice the Chebyshev iteration is implemented as an adaptive scheme, which usually implies a restart after a moderate number of iterations. It should be mentioned that the error-estimation for the CG scheme may also suffer from instabilities. This is mainly due to the loss of orthogonality in the CG process and the possibility of cancelation in (2.12).

Finally, the upper bounds (dashed curve) and lower bounds (dotted curve) for $||\varepsilon_k||_A$ by Dahlquist, Golub and Nash [2] and $||\varepsilon_k||_A$ (continuous curve) are graphed in the next figure for the CG method and $N = 2500$. Again, we used the best possible bounds $\lambda_{min}(A)$ and $\lambda_{max}(A)$ for the spectrum of $A$. Note that, according to Table 3.2., the curve $(||\varepsilon_k^{est}||_A)$ based on the use of continued fractions would not be distinguishable (for this scaling) from the continuous curve.

**Acknowledgement** The first author would like to thank Roland Freund and Henk van der Vorst for several helpful discussions.

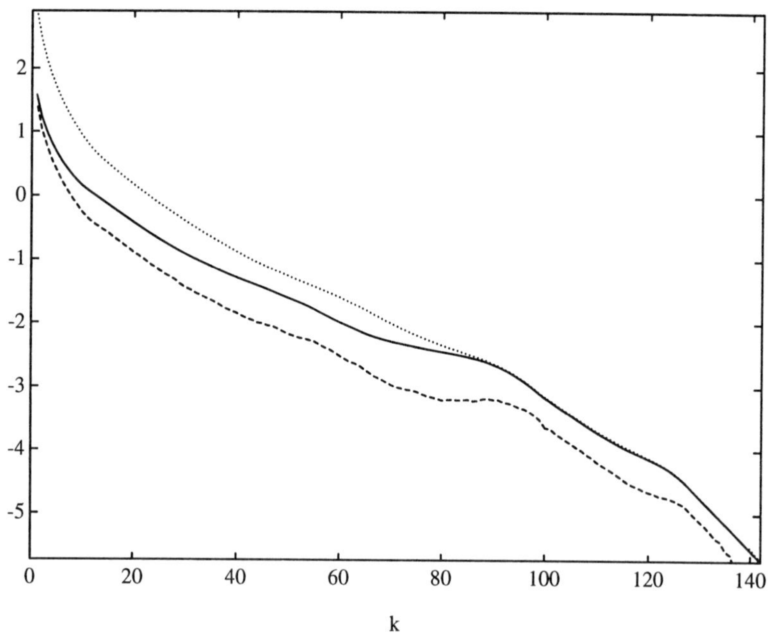


FIG. 3.2. *Estimates of* $||\varepsilon_k||_A$

## REFERENCES

[1] D. Calvetti and L. Reichel, *Adaptive Richardson iteration based on Leja points*, Tech. Rep. ICM-9210-42, Institute for Computational Mathematics, Kent State University, 1992.

[2] G. Dahlquist, G. H. Golub, and S. Nash, *Bounds for the error in linear systems*, in Proceedings of the Workshop on Semi-Infinite Programming, R. Hettich, ed., Springer, 1978, pp. 154–172.

[3] G. G. Dahlquist, S. C. Eisenstat, and G. H. Golub, *Bounds for the error of linear systems of equations using the theory of modified moments*, J. Math. Anal. Appl., 37 (1972), pp. 151–166.

[4] B. Fischer and G. H. Golub, *How to generate unknown orthogonal polynomials out of known orthogonal polynomials*, J. Comp. Appl. Math., 43 (1992), pp. 99–115.

[5] W. Gautschi, *Computational aspects of three-term recurrence relations*, SIAM Rev., 9 (1967), pp. 24–82.

[6] ———, *An algorithmic implementation of the generalized Christoffel theorem*, in Numerical Integration, G. Hämmerlin, ed., Basel, 1982, Birkhäuser, pp. 89–106. Internat. Ser. Numer. Math., v. 57.

[7] G. H. Golub and M. D. Kent, *Estimates of eigenvalues for iterative methods*, Math. Comp., 53 (89), pp. 619–626.

[8] G. H. Golub and C. F. van Loan, *Matrix Computations*, The Johns Hopkins University Press, Baltimore, second ed., 1989.

[9] L. A. Hageman and D. M. Young, *Applied iterative methods*, Academic Press, New York, 1981.

[10] R. A. Sack and A. F. Donovan, *An algorithm for Gaussian quadrature given modified moments*, Numer. Math., 18 (1971/72), pp. 465–478.

[11] J. C. Wheeler, *Modified moments and Gaussian quadrature*, Rocky Mt. J. Math., 4 (1974), pp. 287–296.

[12] H. Wilf, *Mathematics for the Physical Sciences*, Wiley, New York, 1962.

# TRANSPOSE-FREE QUASI-MINIMAL RESIDUAL METHODS FOR NON-HERMITIAN LINEAR SYSTEMS

ROLAND W. FREUND*

**Abstract.** Recently, Freund and Nachtigal proposed a novel conjugate gradient-type method, the quasi-minimal residual algorithm (QMR), for the iterative solution of general non-Hermitian systems of linear equations. The QMR method is based on the nonsymmetric Lanczos process, and thus, like the latter, QMR requires matrix-vector multiplications with both the coefficient matrix of the linear system and its transpose. However, in certain applications, the transpose is not readily available, and generally, it is desirable to trade in multiplications with the transpose for matrix-vector products with the original matrix.

This paper gives a survey of transpose-free algorithms that are based on the quasi-minimal residual approach. First, it is shown that, in principle, the transpose in the standard QMR method can always be eliminated by choosing special starting vectors. Examples are given for which this approach is practical. Then, two transpose-free QMR methods, the TFQMR algorithm and the QMR squared algorithm, for general non-Hermitian systems are described. Some theory for ideal transpose-free QMR and TFQMR is presented. Results of numerical experiments are reported. Finally, some open problems are mentioned.

**Key words.** non-Hermitian linear systems, nonsymmetric Lanczos process, quasi-minimal residual property, transpose-free, conjugate gradients squared, symmetric indefinite linear systems

**AMS(MOS) subject classifications.** 65F10, 65N22, 65K05

## 1. Introduction.

The classical conjugate gradient method (CG) [27]—used in conjunction with preconditioning [5], [30]—is one of the most powerful iterative techniques for solving Hermitian positive definite systems of linear equations. Among the many properties of CG, the following two are especially important:

(i) The method generates iterates that are characterized by a minimization property over Krylov subspaces generated by the coefficient of the linear system.

(ii) The iterates are computable by means of short vector updates, and thus storage requirements and work per iteration are low and constant throughout the iteration.

For general non-Hermitian linear systems, the situation is less satisfactory. It was shown [8], [29], [39] that ideal non-Hermitian extensions of CG with both properties (i) and (ii) do not exist. Instead, most CG-type methods

* AT&T Bell Laboratories, Room 2C–420, 600 Mountain Avenue, Murray Hill, New Jersey 07974–0636. This research was performed while the author was in residence at the Research Institute for Advanced Computer Science (RIACS), NASA Ames Research Center, Moffett Field, California 94035, and it was supported by Cooperative Agreement NCC 2-387 between the National Aeronautics and Space Administration and the Universities Space Research Association.

for non-Hermitian linear systems fulfill either (i) or (ii), but not both. In the first category, the most successful scheme is the generalized minimal residual algorithm (GMRES) [36]. It generates iterates defined by a minimal residual property, and thus it fulfills (i). However, the condition (ii) is violated, since storage requirements and work per iteration grow linearly with the iteration number. In the second category, the archetype is the biconjugate gradient algorithm (BCG) [32], [9]. BCG is based on simple coupled two-term recurrences, and thus the condition (ii) is satisfied. However, the BCG iterates are defined by a Galerkin condition rather than a minimization property (i), which means that the algorithm can exhibit—and typically does—a rather irregular convergence behavior with wild oscillations in the residual norm.

Although one cannot fulfill (i) and (ii) simultaneously, it is possible to devise CG-type methods for general non-Hermitian linear systems that satisfy (ii) and a relaxed version of (i), namely a quasi-minimization property. An iterative scheme of this type, the quasi-minimal residual algorithm (QMR), was recently proposed by Freund [11] for the special case of complex symmetric matrices, and by Freund and Nachtigal [18] for the case of general nonsingular non-Hermitian linear systems. Freund and Hochbruck [17] extended the QMR method to linear systems with singular square coefficient matrices.

The QMR method is based on the nonsymmetric Lanczos process, and thus, like the latter, QMR requires matrix-vector multiplications with both the coefficient matrix $A$ of the linear system and its transpose $A^T$. This is a disadvantage in certain applications where products $A \cdot v$ can be obtained efficiently without explicitly generating $A$, while products $A^T \cdot w$ are not readily available. Furthermore, in QMR, the transpose $A^T$ only appears in the construction of vectors that are used to compute the recurrence coefficients for the Lanczos process. However, these vectors are never used directly in the construction of the actual QMR iterates. Therefore, it is desirable to have QMR methods that do not involve the transpose. Various such transpose-free QMR algorithms have been proposed, see [12], [4], [20], [22]. In this paper, we give an overview of these different transpose-free QMR approaches, and we present some extensions of the results given in [12].

The remainder of this paper is organized as follows. In § 2, we briefly review the standard QMR method. In § 3, we consider the ideal transpose-free QMR algorithm, and we discuss some cases where this approach is viable. In § 4, we describe the TFQMR method. In § 5, we give convergence results for QMR and TFQMR. In § 6, we present an algorithm for computing the zeros of the QMR and TFQMR residual polynomials. In § 7, we describe the QMR squared algorithm. In § 8, results of numerical experiments are reported Finally, in § 9, we make some concluding remarks, and we mention some open problems.

Throughout the paper, all vectors and matrices are allowed to have real or complex entries. As usual, $M^T = [m_{kj}]$ and $M^H = [\overline{m_{kj}}]$ denote the transpose and the conjugate transpose, respectively, of the matrix $M = [m_{jk}]$. The vector norm $\|x\| := \sqrt{x^H x}$ is always the Euclidean norm, and $\|M\| := \max_{\|x\|=1} \|Mx\|$ is the corresponding matrix norm. The set of eigenvalues of a square matrix $M$ is denoted by $\lambda(M)$. We denote by

$$\mathcal{P}_n := \left\{\varphi(\lambda) \equiv \gamma_0 + \gamma_1\lambda + \cdots + \gamma_n\lambda^n \;\middle|\; \gamma_0, \gamma_1, \cdots, \gamma_n \in \mathbb{C}\right\}$$

the set of all complex polynomials of degree at most $n$. The $n$th Krylov subspace of $\mathbb{C}^N$ generated by $c \in \mathbb{C}^N$ and $M \in \mathbb{C}^{N\times N}$ is defined by

$$K_n(c, M) := \operatorname{span}\{c, Mc, \cdots, M^{n-1}c\},$$

and we will make use of the fact that

$$K_n(c, M) = \{\varphi(M)c \mid \varphi \in \mathcal{P}_{n-1}\}.$$

**2. Review of the standard QMR method.** We are concerned with the iterative solution of systems of linear equations

$$Ax = b, \tag{2.1}$$

where $A$ is a general, singular or nonsingular, $N \times N$ matrix with real or complex entries, and $b \in \mathbb{C}^N$. In this section, we briefly review the standard QMR method [18].

**2.1. Krylov subspace iterations.** All algorithms described in this paper fall into the category of *Krylov subspace iterations*: they produce approximate solutions $x_n$ of (2.1) of the form

$$x_n \in x_0 + K_n(r_0, A), \quad n = 1, 2, \cdots. \tag{2.2}$$

Here $x_0 \in \mathbb{C}^N$ is an arbitrary initial guess, and $r_0 := b - Ax_0$ is the corresponding residual vector. In the sequel, we always use the notation

$$r_n := b - Ax_n$$

for the residual vector associated with the iterate $x_n$.

The standard QMR method is a Krylov subspace iteration, where a look-ahead Lanczos algorithm is used to generate basis vectors for the subspaces $K_n(r_0, A)$, and the iterates $x_n$ are characterized by a quasi-minimal residual property.

**2.2. The look-ahead Lanczos algorithm.** In the classical nonsymmetric Lanczos algorithm [31], breakdowns—more precisely, division by 0—can occur. Parlett, Taylor, and Liu [34] were the first to demonstrate that breakdowns and near-breakdowns (division by a nonzero, but small number) in the standard algorithm can be avoided by using so-called *look-ahead* techniques. Recently, Freund, Gutknecht, and Nachtigal [16] developed a look-ahead Lanczos algorithm that is different from the procedure described in [34]. The QMR method is based on the algorithm proposed in [16]. Next we sketch some key properties of this look-ahead Lanczos procedure; a detailed description of the algorithm can be found in [16]. For further properties of the Lanczos process, we refer the reader to [26], [15], [21] and the references given there.

Starting with arbitrary initial vectors

$$v_1,\, w_1 \in \mathbb{C}^N, \quad \text{with} \quad \|v_1\| = \|w_1\| = 1, \tag{2.3}$$

the look-ahead Lanczos algorithm generates two sequences of vectors $v_1, v_2 \cdots$, and $w_1, w_2, \cdots$, such that, for $n = 1, 2, \cdots$,

$$\begin{aligned} \operatorname{span}\{v_1, v_2, \cdots, v_n\} &= K_n(v_1, A), \\ \operatorname{span}\{w_1, w_2, \cdots, w_n\} &= K_n(w_1, A^T), \end{aligned} \tag{2.4}$$

and the two sequences are block biorthogonal:

$$(W^{(j)})^T V^{(k)} = 0 \quad \text{for all} \quad j \neq k, \quad j, k = 1, 2, \cdots, l(n). \tag{2.5}$$

The matrices $W^{(j)}$ and $V^{(k)}$ in (2.5) are defined by partitioning the first $n$ Lanczos vectors $\{v_j\}_{j=1}^n$ and $\{w_j\}_{j=1}^n$ into blocks, according to the look-ahead steps taken:

$$\begin{aligned} V_n := [\, v_1 \;\; v_2 \;\; \cdots \;\; v_n \,] &= [\, V^{(1)} \;\; V^{(2)} \;\; \ldots \;\; V^{(l(n))} \,], \\ W_n := [\, w_1 \;\; w_2 \;\; \cdots \;\; w_n \,] &= [\, W^{(1)} \;\; W^{(2)} \;\; \ldots \;\; W^{(l(n))} \,]. \end{aligned} \tag{2.6}$$

More precisely, the matrices $V^{(k)}$ and $W^{(k)}$ are both of order $N \times \mu_k$, and their columns are just the Lanczos vectors corresponding to the $k$th look-ahead step. Here, $\mu_k$ denotes the length of the $k$th look-ahead step. Moreover, in (2.6), $l(n)$ denotes the number of look-ahead steps that have been performed during the first $n$ iterations of the Lanczos process. We remark that the conditions (2.4)–(2.5) determine the Lanczos vectors only up to scaling. In the sequel, we always assume that the Lanczos vectors are scaled to have unit length:

$$\|v_n\| = \|w_n\| = 1, \quad n = 1, 2, \cdots. \tag{2.7}$$

We stress that the algorithm performs mostly standard Lanczos steps, i.e., look-ahead steps of size $\mu_k = 1$ with blocks $V^{(k)}$ and $W^{(k)}$ that consist of only single Lanczos vectors. True look-ahead steps, i.e., steps of size $\mu_k > 1$, are only used to avoid exact and near-breakdowns. Typically, except for contrived examples, only few true look-ahead steps occur, and their size is usually small, mostly $\mu_k = 2$. We note that, if only steps of length $\mu_k = 1$ are performed, then the look-ahead Lanczos algorithm reduces to the classical algorithm.

The crucial point of the look-ahead Lanczos algorithm is that the vectors satisfying (2.4) and (2.5) can be generated by means of short recurrences. For the first $n+1$ vectors $\{v_j\}_{j=1}^{n+1}$ and $\{w_j\}_{j=1}^{n+1}$, these recurrence relations can be written compactly in matrix form as follows:

$$AV_n = V_{n+1}T_n \quad \text{and} \quad A^T W_n = W_{n+1}\Gamma_{n+1}^{-1}T_n\Gamma_n. \tag{2.8}$$

Here $T_n$ is an $(n+1)\times n$ upper Hessenberg matrix, which is also block tridiagonal with square blocks of order $\mu_k$ on the diagonal, and

$$\Gamma_n := \operatorname{diag}(\gamma_1, \gamma_2, \cdots, \gamma_n), \quad \text{with} \quad \gamma_j > 0 \quad \text{for all} \quad j, \tag{2.9}$$

is a scaling matrix. Furthermore, we note that $T_n$ always has full column rank, and thus

$$\operatorname{rank} T_n = n. \tag{2.10}$$

Finally, we will make use of the fact that each pair of Lanczos vectors $v_n$ and $w_n$ can be expressed in the form

$$v_n = \varphi_{n-1}(A)v_1 \quad \text{and} \quad w_n = \gamma_n\varphi_{n-1}(A^T)w_1, \tag{2.11}$$

where $\varphi_{n-1} \in \mathcal{P}_{n-1}$ is of exact degree $n-1$ and $\gamma_n > 0$ is as in (2.9). We refer to the $\varphi_j$'s in (2.11) as the *look-ahead Lanczos polynomials*.

**2.3. The QMR iterates.** In (2.3), we now choose the starting vector

$$v_1 := r_0/\rho_1, \quad \text{where} \quad \rho_1 := \|r_0\|, \tag{2.12}$$

while $w_1$ is still an arbitrary unit vector. In view of (2.4) and (2.6), the columns of $V_n$ span the subspace $K_n(r_0, A)$ in (2.2). Hence the $n$th iterate $x_n$ of any Krylov subspace iteration can be written in the form

$$x_n = x_0 + V_n z_n, \quad \text{where} \quad z_n \in \mathbb{C}^n. \tag{2.13}$$

The $n$th iterate of the standard QMR method is given by (2.13) where $z_n \in \mathbb{C}^n$ is chosen as the unique solution of the least-squares problem

$$\|f_n - H_n z_n\| = \min_{z\in\mathbb{C}^n} \|f_n - H_n z\|. \tag{2.14}$$

Here $H_n$ is a rescaled version of the Lanczos matrix $T_n$ from (2.8), namely

$$H_n := \Omega_{n+1}T_n, \tag{2.15}$$

where

$$\Omega_{n+1} := \operatorname{diag}(\omega_1, \omega_2, \cdots, \omega_{n+1}), \quad \text{with} \quad \omega_j > 0 \quad \text{for all} \quad j, \tag{2.16}$$

is a still arbitrary diagonal scaling matrix. Moreover, $f_n$ in (2.14) is given by

$$f_n := \omega_1 \rho_1 \begin{bmatrix} 1 & 0 & \cdots & 0 \end{bmatrix}^T \in \mathbb{R}^{n+1}. \tag{2.17}$$

We remark that, by (2.10) and (2.15)–(2.16), the matrix $H_n$ has full column rank, and this guarantees that $z_n$ is uniquely defined by (2.14).

Using (2.13), (2.12), the first relation in (2.8), (2.15), and (2.17), one readily verifies that the residual vector $r_n$ corresponding to $x_n$ satisfies

$$r_n = V_{n+1} \Omega_{n+1}^{-1} (f_n - H_n z_n). \tag{2.18}$$

Hence, in view of (2.14), the $n$th QMR iterate is characterized by a minimization of the second factor in (2.18); this is just the *quasi-minimal residual property*. We note that the standard choice for the scaling matrix (2.16) is

$$\omega_j = 1 \quad \text{for all} \quad j. \tag{2.19}$$

In view of (2.18) and (2.7), the scaling (2.19) is very natural, in the sense that all columns of the matrix $V_{n+1}\Omega_{n+1}^{-1}$ in the representation (2.18) of $r_n$ are treated equally. We stress that there are situations (see § 7.2) where it is crucial to use a scaling different from (2.19); this is the reason for formulating the QMR method for general scaling matrices (2.16).

The original implementation [18] of the QMR method is directly based on the definition (2.13)–(2.14) of the QMR iterates. We remark that, due to the block tridiagonal structure of $H_n$, the solution $z_n$ of (2.14), and hence the QMR iterate $x_n$ can be easily updated from step to step. The resulting update for $x_n$ only involves short vector recurrences; see [18] for details.

Of course, there are also other possible implementations of the QMR method. Suppose that we have a factorization

$$H_n = L_n U_n \tag{2.20}$$

of $H_n$ into an upper Hessenberg matrix $L_n$ and an upper triangular matrix $U_n$. Then, the definition (2.13)–(2.14) of the $n$th QMR iterate can be rewritten as follows:

$$x_n = x_0 + P_n y_n, \tag{2.21}$$

where $P_n := V_n U_n^{-1}$ and $y_n$ $(:= U_n z_n)$ is the unique solution of

$$\|f_n - L_n y_n\| = \min_{y \in \mathbb{C}^n} \|f_n - L_n y\|. \tag{2.22}$$

An implementation of QMR based on (2.21)–(2.22) was recently proposed by Freund and Nachtigal [19]. Here, instead of (2.8), the Lanczos vectors $\{v_j\}_{j=1}^{n+1}$ are generated by coupled recurrences of the form

$$V_n = P_n U_n, \quad AP_n = V_{n+1}\Omega_{n+1}^{-1} L_n, \tag{2.23}$$

and analogous recursions are used to compute $\{w_j\}_{j=1}^{n+1}$. If no look-ahead steps are taken in this coupled procedure, then the matrices $L_n$ and $U_n$ are lower and upper bidiagonal matrices, respectively. For a detailed description of this implementation of QMR based on coupled recurrences, we refer the reader to [19].

**3. The ideal transpose-free QMR method.** We now turn to transpose-free QMR iterations for solving linear systems (2.1). Note that, in the standard QMR method, the transpose $A^T$ only appears in the recurrences for the vectors $\{w_j\}_{j=1}^{n+1}$ in (2.8). These vectors are needed for generating the entries of the Lanczos recurrence matrix $T_n$; however, they are not used in the actual computation of the QMR iterates based on (2.13)–(2.14). Freund and Zha [22] have shown that, in principle, it is always possible to eliminate the transpose in the standard QMR method by choosing a special starting vector $w_1$. In this section, we describe the resulting *ideal* transpose-free QMR method.

**3.1. Simplification of the Lanczos process.** Suppose that we are given a nonsingular matrix $S \in \mathbb{C}^{N\times N}$ such that

$$A^T S = SA. \tag{3.1}$$

We remark that a matrix $S$ satisfying (3.1) always exists, since any square matrix $A$ is similar to its transpose $A^T$ (see, e.g., [28, Chap. 3.2]). Recall that, for the standard QMR method, the second starting vector $w_1$ in (2.3) can be chosen arbitrarily. We now set

$$w_1 := \frac{1}{\|Sv_1\|} Sv_1. \tag{3.2}$$

Then, using (3.1)–(3.2), the polynomial representation (2.11) of $v_n$ and $w_n$, and the normalization condition (2.7), one readily verifies that

$$w_n = \frac{\gamma_n}{\|Sv_1\|} Sv_n = \frac{1}{\|Sv_n\|} Sv_n \quad \text{for all} \quad n. \tag{3.3}$$

Hence, for the special starting vector (3.2), the look-ahead Lanczos algorithm simplifies: instead of running both recurrences in (2.8) for both Lanczos sequences $v_1, v_2, \cdots$, and $w_1, w_2, \cdots$, only the first one is used to generate the $v_n$'s, while the $w_n$'s are computed by means of (3.3). Clearly, this eliminates the need of $A^T$, at the expense of one matrix-vector product $S \cdot v_n$ at each iteration step. Therefore, this ideal transpose-free QMR method is only viable if the matrix $S$ is such that products with $S$ can be computed cheaply. Examples of this type are given in § 3.3.

**3.2. An implementation without look-ahead.** In this section, we present an actual implementation of the ideal transpose-free QMR method. For simplicity, we only state the algorithm without look-ahead; the general case with look-ahead is treated in [22]. Moreover, in practical applications, the matrix $S$ in (3.1) usually corresponds to the inverse of a preconditioner $M$, and we formulate the algorithm in this setting.

Let $M \in \mathbb{C}^{N\times N}$ be an arbitrary given nonsingular matrix. We then define

$$J := SM \quad \text{and} \quad T := AM, \tag{3.4}$$

and we state the ideal transpose-free QMR algorithm in terms of the matrices $T$, $M$, and $J$, instead of $A$ and $S$. We remark that, with (3.4), the relation (3.1) for $S$ translates into the equivalent condition

$$T^T J = M^T J M^{-1} T \tag{3.5}$$

for $J$. The motivation for the transition to $T$, instead of $A$, is as follows. Suppose we want to solve the linear system

$$Ty = b, \tag{3.6}$$

using the matrix $M$ as a right preconditioner. We apply the ideal transpose-free QMR method to the preconditioned system

$$Ax = b, \quad \text{where} \quad A := TM^{-1} \quad \text{and} \quad x := My.$$

By rewriting the resulting algorithm in terms of the original linear system (3.6), one obtains the following ideal transpose-free QMR algorithm for solving (3.6), using a right preconditioning matrix $M$.

ALGORITHM 3.1. ("IDEAL" TRANSPOSE-FREE QMR WITHOUT LOOK-AHEAD)

1. *Start:*
   (a) *Choose* $y_0 \in \mathbb{C}^N$.
   (b) *Set* $r_0 = b - Ty_0$, $\rho_1 = \|r_0\|$, $v_1 = r_0/\rho_1$, $p_0 = d_0 = 0$;
   $\chi_0 = \varepsilon_0 = 1$, $\vartheta_0 = 0$, $\eta_0 = -1$.
2. *For* $n = 1, 2, \cdots,$ *do:*
   (a) *Compute* $\hat{w}_n$ *by solving* $M\hat{w}_n = v_n$.
   *Set* $\delta_n = (J\hat{w}_n)^T v_n$, $\beta_{n-1} = \rho_n \delta_n / \varepsilon_{n-1}$;
   $p_n = \hat{w}_n - p_{n-1}\beta_{n-1}$.
   (b) *Set* $\varepsilon_n = (Jp_n)^T T p_n$, $\alpha_n = \varepsilon_n/\delta_n$;
   $\tilde{v}_{n+1} = Tp_n - v_n\alpha_n$, $\rho_{n+1} = \|\tilde{v}_{n+1}\|$, $v_{n+1} = \tilde{v}_{n+1}/\rho_{n+1}$.
   (c) *Set* $\vartheta_n = (\omega_{n+1}\rho_{n+1})^2/(\omega_n^2 \chi_{n-1}|\alpha_n|^2)$, $\chi_n = 1/(1+\vartheta_n)$;
   $\eta_n = -\eta_{n-1}\rho_n\chi_n/(\alpha_n\chi_{n-1})$;
   $d_n = p_n\eta_n + d_{n-1}(\vartheta_{n-1}\chi_n)$, $y_n = y_{n-1} + d_n$.
   
   *If* $y_n$ *has converged, then stop.*

Note that, in Algorithm 3.1, we only generate the vectors $\hat{w}_n := M^{-1}v_n$ ($= J^{-1}Sv_n$), instead of (3.3). The computation of $\hat{w}_n$ involves the usual solution of a linear system with the preconditioner $M$ as coefficient matrix. In addition, Algorithm 3.1 involves two matrix-vector products $J \cdot \hat{w}_n$ and $J \cdot p_n$ at each step. In typical applications (see § 3.3), the matrix $J$ is either the identity or the antidiagonal identity matrix, and then these multiplications with $J$ do not represent extra work.

Finally, we remark that Algorithm 3.1 uses coupled recurrences to generate the Lanczos vectors $\{v_j\}_{j=1}^{n+1}$, based on a factorization (2.20) of $H_n$ into a lower and upper bidiagonal matrix $L_n$ and $U_n$, respectively. For details of the derivation of Algorithm 3.1, we refer the reader to [22], [19].

**3.3. Examples.** In this section, we give examples for linear systems (3.6) with preconditioning matrices $M$ that satisfy (3.5).

**3.3.1. Complex symmetric systems.** Many important applications lead to linear systems (3.6) with complex non-Hermitian, but symmetric coefficient matrices $T = T^T$; we refer the reader to [11] and the references given there. Standard preconditioning techniques, as for instance incomplete factorization [30], yield complex symmetric preconditioners $M = M^T$ when applied to complex symmetric systems. For complex symmetric $T$ and $M$, the condition (3.5) is trivially satisfied, where $J := I_N$ is the $N \times N$ identity matrix.

**3.3.2. Real symmetric indefinite systems.** Another important special case is real linear systems (3.6) where $T = T^T \in \mathbb{R}^{N \times N}$ is symmetric, but indefinite and $M = M^T \in \mathbb{R}^{N \times N}$ is a nonsingular preconditioner for (3.6). Again, condition (3.5) is trivially satisfied with $J := I_N$. The resulting QMR Algorithm 3.1 for solving real symmetric indefinite systems has the same work and storage requirements as SYMMLQ and MINRES [33], which are the standard CG-like iterations for such systems. However, both SYMMLQ and MINRES can only be used with symmetric positive definite preconditioners $M$. This is a rather restrictive condition for highly indefinite systems for which standard preconditioning techniques yield indefinite preconditioners. In contrast to SYMMLQ and MINRES, the QMR Algorithm 3.1 can be combined with any nonsingular symmetric preconditioner $M$.

We remark that a particular class of highly indefinite symmetric systems are real linear systems (3.6) with coefficient matrices of the form

$$T = \begin{bmatrix} H & G^T \\ G & 0 \end{bmatrix}, \tag{3.7}$$

where $H$ is symmetric positive semidefinite. Such systems arise, for example, in incompressible flow computations [25] and as subproblems in optimization algorithms, such as interior-point methods for linear and nonlinear programs or sequential programming methods for constrained nonlinear programs [23], [40].

**3.3.3. Centrosymmetric systems.** A matrix $T \in \mathbb{C}^{N\times N}$ is called *centrosymmetric* if it is symmetric with respect to the antidiagonal, i.e.,

$$T = J_N T^T J_N,$$

where $J_N$ is the $N \times N$ antidiagonal identity matrix defined by

$$J_N := \begin{bmatrix} 0 & \cdots & 0 & 1 \\ \vdots & \iddots & 1 & 0 \\ 0 & \iddots & \iddots & \vdots \\ 1 & 0 & \cdots & 0 \end{bmatrix} \in \mathbb{R}^{N\times N}.$$

One easily verifies that, for centrosymmetric linear systems (3.6) with centrosymmetric preconditioners $M$, the condition (3.5) is satisfied with $J := J_N$. Hence the QMR Algorithm 3.1 can be used for the iterative solution of centrosymmetric linear systems. Note that matrix-vector products $J \cdot v$ just correspond to reversing the order of the entries of $v$.

We note that Toeplitz matrices, i.e., matrices whose elements are constant along each diagonal, are in particular centrosymmetric. Chan and Strang [3] suggested to use circulant matrices as preconditioners for Toeplitz systems. In this case, the linear systems with the preconditioner $M$ can be solved efficiently be means of the fast Fourier transformation. Since circulant matrices are centrosymmetric, the QMR Algorithm 3.1 can be used for the iterative solution of Toeplitz systems (3.6) with circulant preconditioners.

Finally, we remark that further examples, for which the ideal transpose-free QMR method is practical, are given in [22].

**4. The TFQMR method.** In this section, we sketch the *transpose-free quasi-minimal residual* method (TFQMR) for solving general non-Hermitian linear systems (2.1).

**4.1. Sonneveld's observation.** Sonneveld [37] with his *conjugate gradients squared* algorithm (CGS) was the first to propose a transpose-free BCG-type Krylov subspace iteration for non-Hermitian linear systems (2.1). All scalar quantities that appear in the $n$th iteration of BCG are of the form

$$w^T v, \quad \text{where} \quad v = \varphi(A)v_1, \; w = \psi(A^T)w_1 \quad \text{and} \quad \varphi, \psi \in \mathcal{P}_n. \tag{4.1}$$

Sonneveld observed that, by rewriting (4.1) in the form

$$w^T v = w_1^T(\psi(A)\varphi(A)v_1),$$

the BCG scalars can be obtained without using $A^T$, by working with vectors of the form $\psi(A)\varphi(A)v_1$ instead of $\varphi(A)v_1$ and $\psi(A^T)w_1$. Although the CGS algorithm produces the BCG scalars, it generates a different set of iterates than BCG. More precisely, the CGS iterates $x_{2n}^{\mathrm{CGS}}$, $n = 0, 1, \cdots$, are defined by

$$r_{2n}^{\mathrm{CGS}} = b - Ax_{2n}^{\mathrm{CGS}} = (\psi_n^{\mathrm{BCG}}(A))^2 r_0. \tag{4.2}$$

Here $\psi_n^{\mathrm{BCG}}$ is given by the polynomial representation

$$\begin{aligned} r_n^{\mathrm{BCG}} = & \quad b - Ax_n^{\mathrm{BCG}} = \psi_n^{\mathrm{BCG}}(A) r_0, \\ & \text{where} \quad \psi_n^{\mathrm{BCG}} \in \mathcal{P}_n \quad \text{and} \quad \psi_n^{\mathrm{BCG}}(0) = 1, \end{aligned} \tag{4.3}$$

of the $n$th BCG residual vector. We remark that, in CGS and BCG, roughly the same work and storage is required to generate the iterates

$$x_{2n}^{\mathrm{CGS}} \in x_0 + K_{2n}(r_0, A) \quad \text{and} \quad x_n^{\mathrm{BCG}} \in x_0 + K_n(r_0, A),$$

respectively.

Due to the squaring of the polynomial in (4.2), the CGS algorithm usually converges faster than BCG. However, the squaring also amplifies the erratic convergence behavior of BCG, and there are cases where BCG converges, while CGS diverges. Finally, breakdowns—triggered by division by 0—can occur in the CGS algorithm. Actually, in exact arithmetic, CGS breaks down in the $n$th step if, and only if, BCG breaks down during the $n$th iteration.

**4.2. The general TFQMR method.** The $n$th CGS iterate $x_{2n}^{\mathrm{CGS}}$ is defined in (4.2) by squaring the BCG polynomial, and in general, it does not satisfy a minimization or quasi-minimization property. In [12], we proposed a transpose-free Krylov subspace iteration, the TFQMR algorithm, that uses the CGS search directions to generate iterates characterized by a quasi-minimal residual property. The TFQMR algorithm in [12] is derived directly from the CGS process, and like the latter, it can break down. In this section, we present an extension of the TFQMR method that is based on the look-ahead Lanczos polynomials, and thus avoids breakdowns.

Instead of one CGS iterate $x_{2n}^{\mathrm{CGS}}$, the TFQMR method produces two iterates

$$x_m \in x_0 + K_m(r_0, A), \quad m = 2n-1,\, 2n, \tag{4.4}$$

in step $n$. Therefore, in the sequel, we will use $m = 1, 2, \cdots$, as the iteration index in our description of TFQMR. First, based on the look-ahead Lanczos polynomials given in (2.11), we define vectors

$$u_m := \begin{cases} \dfrac{\varphi_{n-1}(A)\varphi_n(A)r_0}{\|\varphi_{n-1}(A)\varphi_n(A)r_0\|} & \text{if m=2n,} \\[2ex] \dfrac{(\varphi_n(A))^2 r_0}{(\|\varphi_n(A))^2 r_0\|} & \text{if m=2n+1.} \end{cases} \tag{4.5}$$

Since each polynomial $\varphi_n$ is of exact degree $n$, we have

$$K_m(r_0, A) = \{U_m z \mid z \in \mathbb{C}^m\}, \tag{4.6}$$

where

$$U_m := \begin{bmatrix} u_1 & u_2 & \cdots & u_m \end{bmatrix}. \tag{4.7}$$

Furthermore, the vectors $\{u_j\}_{j=1}^{m+1}$ satisfy recurrences of the form

$$AU_m = U_{m+1} F_m, \tag{4.8}$$

where $F_m$ is an $(m+1) \times m$ upper Hessenberg matrix with nonzero subdiagonal elements. In particular, we have

$$\operatorname{rank} F_m = m. \tag{4.9}$$

Finally, note that, in view of (2.11), $\varphi_0(\lambda) \equiv 1$, and thus, by (4.5), $u_1 = r_0/\rho_1$, where $\rho_1 := \|r_0\|$.

By (4.6), the iterates (4.4) can be parametrized in the form

$$x_m = x_0 + U_m z_m, \quad \text{where} \quad z_m \in \mathbb{C}^m, \tag{4.10}$$

and, with (4.8), the corresponding residual vector $r_m$ satisfies

$$r_m = U_{m+1} \Omega_{m+1}^{-1} (f_m - H_m z_m). \tag{4.11}$$

Here, $\Omega_{m+1}$ is again an arbitrary scaling matrix of the form (2.16), $f_m$ is defined as in (2.17), and

$$H_m := \Omega_{m+1} F_m. \tag{4.12}$$

In analogy to the definition (2.13)–(2.14) of the standard QMR iterates, the $m$th TFQMR iterate is given by (4.10) where $z_m$ is chosen so that the Euclidean norm of the second factor in (4.11) is as small as possible, i.e., $z_m \in \mathbb{C}^m$ is the solution of the least-squares problem

$$\|f_m - H_m z_m\| = \min_{z \in \mathbb{C}^m} \|f_m - H_m z\|. \tag{4.13}$$

Note that, in view of (4.9), $z_m$ is uniquely defined by (4.13). We remark that—analogous to the scaling (2.19) in the case of the QMR method—the standard choice for the weights in $\Omega_{m+1}$ is

$$\omega_j = 1 \quad \text{for all} \quad j, \tag{4.14}$$

so that all columns of the matrix $U_{m+1}\Omega_{m+1}^{-1}$ in the representation (4.11) of $r_m$ are scaled to have unit length. Finally, we would like to stress that the TFQMR method and the standard QMR method are not mathematically equivalent. In particular, the sets of iterates generated by TFQMR and QMR are different.

**4.3. An implementation of TFQMR without look-ahead.** Recall that the basis vectors in (4.5) involve the look-ahead Lanczos polynomials. As a result, any robust implementation of TFQMR requires look-ahead in order to avoid breakdowns or near-breakdowns. It is an open question, whether TFQMR with look-ahead can be implemented with only short vector recurrences and with only two matrix-vector multiplications by $A$ per $n$th iteration. An actual implementation of TFQMR without look-ahead was derived in [12]. In this section, we present this algorithm.

We assume that no breakdowns occur in the CGS algorithm, and we denote by $\{y_j\}_{j=1}^{2n}$ the search directions generated during the first $n$ CGS iterations. In analogy to (4.7), we set

$$Y_m := \begin{bmatrix} y_1 & y_2 & \cdots & y_m \end{bmatrix}.$$

The columns of $Y_m$ and $U_m$ both span $K_m(r_0, A)$, and thus we have $U_m = Y_m C_m$, where $C_m$ is a nonsingular upper triangular matrix. By rewriting (4.10) and (4.13) in terms of $Y_m$, it follows that the $m$th TFQMR iterate is given by

$$x_m = x_0 + Y_m q_m,$$

where $q_m := C_m z_m$ is the unique solution of

$$\|f_m - H_m C_m^{-1} z_m\| = \min_{z \in \mathbf{C}^m} \|f_m - H_m C_m^{-1} z\|. \tag{4.15}$$

It turns out (see [12]) that the matrix $H_m C_m^{-1}$ is lower bidiagonal, and the solution of (4.15) and thus $x_m$ can be easily updated from step to step. Together with the fact that the CGS search directions can be generated by means of short recurrences, it follows that TFQMR without look-ahead can be implemented using only short vector updates.

Of course, in practice, the TFQMR method is combined with preconditioning. Let $M \in \mathbf{C}^{N \times N}$ be a nonsingular preconditioner for the linear system (2.1), and assume that $M$ is given in the form $M = M_1 M_2$. The algorithm is then applied to the preconditioned system

$$A' x' = M_1^{-1} b, \quad \text{where} \quad A' := M_1^{-1} A M_2^{-1} \quad \text{and} \quad x' := M_2 x, \tag{4.16}$$

and it generates approximate solutions $x'_m$ of (4.16). We denote by $x_m := M_2^{-1} x'_m$ the corresponding iterates for the original system (2.1). We remark that usually one chooses $M_1 = I_N$ and $M_2 = M$ (right preconditioning), or $M_1 = M$ and $M_2 = I_N$ (left preconditioning).

Next, we state the resulting preconditioned TFQMR algorithm without look-ahead, for the standard weighting strategy (4.14).

ALGORITHM 4.1. (TFQMR WITHOUT LOOK-AHEAD; WITH WEIGHTS (4.14))

*1. Start:*

(a) *Choose* $x_0 \in \mathbb{C}^N$ *and set* $x_0' = M_2 x_0$.
(b) *Set* $\hat{u}_1 = y_1 = M_1^{-1}b - A'x_0'$, $v_0 = A'y_1$, $d_0 = 0$;
$\tau_0 = ||\hat{u}_1||$, $\vartheta_0 = 0$, $\eta_0 = 0$.
(c) *Choose* $\tilde{r}_0$ *such that* $\rho_0 = \tilde{r}_0^H \hat{u}_1 \neq 0$.

2. *For* $n = 1, 2, \cdots$, *do*:
(a) *Set* $\sigma_{n-1} = \tilde{r}_0^H v_{n-1}$, $\alpha_{n-1} = \rho_{n-1}/\sigma_{n-1}$;
$y_{2n} = y_{2n-1} - \alpha_{n-1} v_{n-1}$.
(b) *For* $m = 2n-1, 2n$ *do*:
   - *Set* $\hat{u}_{m+1} = \hat{u}_m - \alpha_{n-1} A' y_m$;
   - $\vartheta_m = ||\hat{u}_{m+1}||/\tau_{m-1}$, $c_m = 1/\sqrt{1+\vartheta_m^2}$;
   - $\tau_m = \tau_{m-1}\vartheta_m c_m$, $\eta_m = c_m^2 \alpha_{n-1}$;
   - $d_m = y_m + (\vartheta_{m-1}^2 \eta_{m-1}/\alpha_{n-1}) d_{m-1}$;
   - $x_m' = x_{m-1}' + \eta_m d_m$.
   - *If* $x_m'$ *has converged, then set* $x_m = M_2^{-1} x_m'$ *and stop*.

(c) *Set* $\rho_n = \tilde{r}_0^H \hat{u}_{2n+1}$, $\beta_n = \rho_n/\rho_{n-1}$;
$y_{2n+1} = \hat{u}_{2n+1} + \beta_n y_{2n}$;
$v_n = A' y_{2n+1} + \beta_n (A' y_{2n} + \beta_n v_{n-1})$.

We note that the vectors $\hat{u}_m$ in Algorithm 4.1 are rescaled versions of the vectors $u_m$ defined in (4.5). Moreover, the polynomials $\varphi_{n-1}$ and $\varphi_n$ in (4.5) correspond to the classical nonsymmetric Lanczos process without look-ahead.

Finally, we remark that it is straightforward to rewrite Algorithm 4.1 in terms of the quantities $A$, $M_1^{-1}$, $M_2^{-1}$, and $x_m$, instead of $A'$ and $x_m'$. However, in some cases it is more advantageous to work directly with the preconditioned matrix $A'$. For example, this is true for SSOR preconditioning, where products with the preconditioned matrix $A'$ can be computed efficiently using Eisenstat's trick [7].

**5. Convergence results for QMR and TFQMR.** The quasi-minimal residual property of the QMR and TFQMR iterates is strong enough to deduce from it convergence results for both methods, see [18], [14]. In this section, we state these results under some additional weak assumptions. Formulations of the convergence results for the general case can be found in [14] (for nonsingular $A$) and [17] (for singular $A$).

Note that, in our description of QMR and TFQMR, we have included the case of singular coefficient matrices $A$. Indeed, as was shown by Freund and Hochbruck [17], both QMR and TFQMR generate well-defined iterates in the case of singular $A$. However, like other Krylov subspace methods, the iterates converge to a meaningful solution of (2.1) only for special cases, for instance consistent singular systems with one-dimensional null spaces. This is discussed in detail in [17]. In this section, the coefficient matrix $A$ of (2.1) is always assumed to be nonsingular.

We need some further notation. For any $(n+1) \times n$ matrix $H_n$, we denote by

$$\tilde{H}_n = \begin{bmatrix} I_n & 0 \end{bmatrix} H_n \tag{5.1}$$

the matrix obtained from $H_n$ by deleting its last row. In exact arithmetic, the look-ahead Lanczos algorithm sketched in § 2.2 terminates after at most $N$ steps, either because the process has found an invariant subspace or due to an incurable breakdown, see, e.g., [26], [16]. We denote by $L$ ($\leq N$) the termination index of the look-ahead Lanczos algorithm. Let $\lambda_1, \lambda_2, \cdots, \lambda_L$ be the zeros of the last look-ahead Lanczos polynomial $\varphi_L$, and let $\tilde{T}_L$ be the last $L \times L$ Lanczos matrix defined by (2.8) and (5.1). It is well known (see, e.g., [26]) that

$$\{\lambda_1, \lambda_2, \cdots, \lambda_L\} = \lambda(\tilde{T}_L) \subseteq \lambda(A). \tag{5.2}$$

For the remainder of this section, we assume that the zeros of $\varphi_L$ are all distinct. In view of (5.2), this guarantees that the matrix $\tilde{T}_L$ is similar to the diagonal matrix

$$D_T := \operatorname{diag}(\lambda_1, \lambda_2, \cdots, \lambda_L). \tag{5.3}$$

Moreover, we consider the QMR and TFQMR methods only for the standard weighting strategies (2.19) and (4.14), respectively.

After these preliminaries, we can now state the convergence results. First, we consider the standard QMR method.

THEOREM 5.1. *Let $L$ be the termination index of the look-ahead Lanczos algorithm, and assume that the zeros $\lambda_1, \lambda_2, \cdots, \lambda_L$ of $\varphi_L$ are all distinct. Let $D_T$ be the diagonal matrix defined in* (5.3). *Then, for $n = 1, 2, \cdots, L-1$, the residual vectors $r_n$ generated by the standard QMR method with weights* (2.19) *satisfy*:

$$\begin{aligned} \frac{\|r_n\|}{\|r_0\|} \leq c_T \quad & \sqrt{n+1}\, \min\nolimits_{\varphi \in \mathcal{P}_n:\, \varphi(0)=1} \max\nolimits_{\lambda \in \lambda(A)} |\varphi(\lambda)|, \\ & \textit{where} \quad c_T := \min\nolimits_{X:\, X^{-1}\tilde{T}_L X = D_T} \|X\| \cdot \|X^{-1}\|. \end{aligned} \tag{5.4}$$

We remark that—except for the factor $\sqrt{n+1}$, which reflects the quasi-minimality, and a slightly different constant $c_T$—the upper bound (5.4) is the same as the standard estimate [36] for algorithms such as GMRES that are based on a true minimal residual property.

Next, we turn to the TFQMR method. As we pointed out in [13], TFQMR terminates at iteration step $m = 2L$. Let $\tilde{F}_{2L}$ be the corresponding last TFQMR matrix as defined by (4.8) and (5.1). From the results in [14], it follows that

$$\{\lambda_1, \lambda_2, \cdots, \lambda_L\} = \lambda(\tilde{F}_{2L}),$$

where each $\lambda_j$ is an eigenvalue of algebraic multiplicity 2 and of geometric multiplicity 1. Hence $\tilde{F}_{2L}$ has the Jordan canonical form

$$\begin{aligned} J_F := \quad & \operatorname{diag}(J(\lambda_1), J(\lambda_2), \cdots, J(\lambda_L)), \\ & \text{where} \quad J(\lambda_j) := \begin{bmatrix} \lambda_j & 1 \\ 0 & \lambda_j \end{bmatrix} \quad \text{for all} \quad j. \end{aligned} \tag{5.5}$$

The convergence result for TFQMR can now be formulated as follows.

THEOREM 5.2. *Let $L$ be the termination index of the look-ahead Lanczos algorithm, and assume that the zeros $\lambda_1, \lambda_2, \cdots, \lambda_L$ of $\varphi_L$ are all distinct. Let $J_F$ be the Jordan matrix defined in* (5.5). *Then, for $m = 1, 2, \cdots, 2L-1$, the residual vectors $r_m$ generated by the TFQMR method with weights* (4.14) *satisfy*:

$$\frac{\|r_m\|}{\|r_0\|} \le c_F \quad \sqrt{m+1} \min_{\varphi \in \mathcal{P}_m:\, \varphi(0)=1} \max_{\lambda \in \lambda(A)} \left( |\varphi(\lambda)| + |\varphi'(\lambda)| \right),$$

$$\text{where} \quad c_F := \min_{X:\, X^{-1}\tilde{F}_{2L}X = J_F} \|X\| \cdot \|X^{-1}\|. \tag{5.6}$$

Recall that, in the general case, $n$ steps of the standard QMR method require $n$ matrix-vector multiplications with $A$ and $n$ multiplications with $A^T$, while $m$ TFQMR iterations involve $m$ matrix-vector products with $A$ only. In other words, with $2n$ matrix-vector multiplications, the QMR and TFQMR methods generate iterates

$$x_n^{\mathrm{QMR}} \in x_0 + K_n(r_0, A) \quad \text{and} \quad x_{2n}^{\mathrm{TFQMR}} \in x_0 + K_{2n}(r_0, A),$$

respectively. However, as the presence of the term $|\varphi'(\lambda)|$ in the upper bound (5.6) indicates, $x_{2n}^{\mathrm{TFQMR}}$ is not an optimal quasi-minimal residual iterate in the shifted Krylov subspace $x_0 + K_{2n}(r_0, A)$. Note that, in view of Theorem 5.1, such an optimal quasi-minimal residual iterate would satisfy an estimate of the type (5.4) (with $n$ replaced by $2n$).

## 6. Computing zeros of residual polynomials.

The residual vector $r_n$ corresponding to the $n$th iterate of any Krylov subspace method can be expressed in terms of polynomials:

$$r_n = \psi_n(A) r_0, \quad \text{where} \quad \psi_n \in \mathcal{P}_n \quad \text{and} \quad \psi_n(0) = 1.$$

In general, any $\psi_n \in \mathcal{P}_n$ with $\psi_n(0) = 1$ is called an $n$th *residual polynomial*. It is well known (see, e.g., [13]) that minimal residual algorithms, such as GMRES, are based on special residual polynomials, namely so-called *kernel polynomials* [38]. In [13], we introduced the more general concept of *quasi-kernel polynomials*, and we showed that both QMR and TFQMR are based on particular instances of quasi-kernel polynomials. The zeros of the GMRES, QMR, and TFQMR residual polynomials yield approximate eigenvalues of $A$ that—in the context of solving linear systems—are often more appropriate than the usual *Ritz values* obtained from the Arnoldi and Lanczos processes. In order to distinguish the two sets of approximate eigenvalues, we refer to the zeros of residual polynomials as *pseudo-Ritz values*. We remark that the notation pseudo-Ritz values was first introduced in [10] for the zeros of kernel polynomial associated with Hermitian indefinite linear systems.

In this section, we describe an algorithm for computing the GMRES, QMR, and TFQMR pseudo-Ritz values. First, we recall the following result.

THEOREM 6.1. ([13, COR. 5.3]) *The nth QMR (with weights (2.19)), TFQMR (with weights (4.14)), and GMRES residual polynomials are given by*

$$\psi_n(\lambda) \equiv \frac{\det\left(H_n^H H_n - \lambda \tilde{H}_n^H\right)}{\det\left(H_n^H H_n\right)}.$$

*Here, in the case of QMR and TFQMR, $H_n$ is defined in (2.15) and (4.12) (both with $\Omega_{n+1} := I_{n+1}$), respectively; and in the case of GMRES, $H_n$ is the $(n+1)\times n$ upper Hessenberg matrix generated by n steps of the Arnoldi process (see, e.g., [36]). In all three cases, $\tilde{H}_n$ is the $n \times n$ matrix defined by (5.1).*

In view of Theorem 6.1, the QMR, TFQMR, and GMRES pseudo-Ritz values can be computed via solution of a generalized eigenvalue problem of the form

$$H_n^H H_n z = \lambda \tilde{H}_n^H z, \quad z \in \mathbb{C}^n, \ z \neq 0. \tag{6.1}$$

However, for numerical reasons, it is advisable to avoid forming the matrix $H_n^H H_n$ in (6.1). Instead, as described in [13], one rewrites (6.1) using a QR decomposition of $H_n$ of the form

$$Q_n H_n = \begin{bmatrix} R_n \\ 0 \end{bmatrix}. \tag{6.2}$$

Here, $R_n$ is an upper triangular matrix, and $Q_n$ is a unitary matrix of the form

$$Q_n = G_n \begin{bmatrix} G_{n-1} & 0 \\ 0 & 1 \end{bmatrix} \cdot \begin{bmatrix} G_{n-2} & 0 \\ 0 & I_2 \end{bmatrix} \cdots \begin{bmatrix} G_1 & 0 \\ 0 & I_{n-1} \end{bmatrix}, \tag{6.3}$$

where, for $j = 1, 2, \cdots, n$,

$$G_j = \begin{bmatrix} I_{j-1} & 0 & 0 \\ 0 & c_j & -\overline{s_j} \\ 0 & s_j & c_j \end{bmatrix}, \quad \text{with} \quad c_j \in \mathbb{R}, \ s_j \in \mathbb{C}, \ c_j^2 + |s_j|^2 = 1. \tag{6.4}$$

With (6.2), the generalized eigenvalue problem (6.1) can be reformulated as follows:

$$R_n z = \lambda \tilde{Q}_n z, \quad z \in \mathbb{C}^n, \ z \neq 0, \tag{6.5}$$

where

$$\tilde{Q}_n = \begin{bmatrix} I_{n-1} & 0 \\ 0 & c_n \end{bmatrix} G_{n-1} \begin{bmatrix} G_{n-2} & 0 \\ 0 & 1 \end{bmatrix} \cdots \begin{bmatrix} G_1 & 0 \\ 0 & I_{n-2} \end{bmatrix}. \tag{6.6}$$

Hence, we have the following algorithm for computing pseudo-Ritz values.

ALGORITHM 6.1. (FOR COMPUTING QMR, TFQMR, OR GMRES PSEUDO-RITZ VALUES)

*Input*: *Let* $H_n$ *be the* $(n+1)\times n$ *Hessenberg matrix given by by* (2.15) *with* $\Omega_{n+1} = I_{n+1}$ *(for QMR), by* (4.12) *with* $\Omega_{n+1} = I_{n+1}$ *(for TFQMR), or by the Arnoldi process (for GMRES).*

(a) *Compute the QR decomposition* (6.2) *of* $H_n$ *with* $Q_n$ *of the form* (6.3)–(6.4).

(b) *Compute* $\tilde{Q}_n$ *using* (6.6).

(c) *Apply the QZ algorithm (see, e.g., [24, Chap. 7.7]) to the generalized eigenvalue problem* (6.5). *The resulting eigenvalues are the pseudo-Ritz values corresponding to QMR, TFQMR, or GMRES, respectively.*

**7. The QMR squared algorithm.** In this section, we discuss yet another transpose-free QMR approach based on "squaring" the standard QMR method, in analogy to the derivation of the CGS residuals (4.2) by squaring the BCG residuals (4.3).

**7.1. The basic approach.** For the moment, we assume that the weights (2.16) are arbitrary, but fixed. Let $x_n^{\mathrm{QMR}}$ be the $n$th QMR iterate given by (2.13)–(2.14), corresponding to these weights. Let $\psi_n \in \mathcal{P}_n$ with $\psi_n(0) = 1$ be the associated residual polynomial:

$$r_n^{\mathrm{QMR}} = b - Ax_n^{\mathrm{QMR}} = \psi_n(A)r_0. \tag{7.1}$$

We define the $n$th iterate $x_{2n}$ of the *quasi-minimal residual squared* algorithm (QMRS) by squaring the relation (7.1). Thus we have,

$$r_{2n} = b - Ax_{2n} = \left(\psi_n(A)\right)^2 r_0,$$

and the iterate itself is given by

$$x_{2n} = x_0 + \phi_{2n}(A)r_0, \quad \text{where} \quad \phi_{2n}(\lambda) \equiv \frac{1-\psi_n^2(\lambda)}{\lambda}. \tag{7.2}$$

Of course, it remains to give an actual algorithm for computing the QMRS iterate defined in (7.2). It is an open question whether there exists an implementation of QMRS with look-ahead that uses only short vector updates and only two matrix-vector products per $n$th step.

**7.2. An implementation of QMRS without look-ahead.** In [20], Freund and Szeto derived an implementation of QMRS without look-ahead, and we now briefly describe this algorithm. The key observation here is that the QMR and BCG residuals are connected by a relation of the form

$$\begin{aligned} r_n^{\mathrm{QMR}} &= \mu_n r_{n-1}^{\mathrm{QMR}} + \nu_n r_n^{\mathrm{BCG}}, \\ &\text{where} \quad 0 \le \mu_n \le 1 \quad \text{and} \quad \nu_n := 1-\mu_n, \end{aligned} \tag{7.3}$$

which is due to Freund and Nachtigal [18, Prop. 4.1]. By rewriting (7.3), by means of (4.3) and (7.1), in terms of residual polynomials, we obtain

$$\psi_n(\lambda) \equiv \mu_n \psi_{n-1}(\lambda) + \nu_n \psi_n^{\mathrm{BCG}}(\lambda). \tag{7.4}$$

By squaring the relation (7.4) and by using (7.2) and (4.2), one easily verifies that the QMRS iterates satisfy

$$\begin{aligned} x_{2n} = {} & \mu_n^2 x_{2n-2} + \nu_n^2 x_{2n}^{\mathrm{CGS}} + 2\mu_n \nu_n f_n, \\ & \text{where} \quad f_n := x_0 + A^{-1}(I_N - \psi_{n-1}(A)\psi_n^{\mathrm{BCG}}(A))r_0. \end{aligned} \tag{7.5}$$

It turns out (see [20] for details) that the auxiliary vector $f_n$ can be updated with short vector recursions, at the expense of one extra matrix-vector multiplication with $A$ at each iteration. Recall from [37] that the CGS iterates $x_{2n}^{\mathrm{CGS}}$ can be generated by means of short vector updates and two matrix-vector products with $A$ per step. Thus, in view of (7.5), the computation of the QMRS iterates $x_{2n}$ involves only short vector updates, but three matrix-vector products with $A$ per $n$th iteration. Note that both CGS and TFQMR involve only two multiplications by $A$ at the $n$th step.

So far, we have not specified how to choose the weights $\omega_j$ in (2.16) for the QMRS algorithm. Ideally, as in the standard QMR method, we would like to use unit weights (2.19). It turns out that this is not possible. Indeed, in [20], we showed that the computation of the recurrence coefficients $\mu_n$ and $\nu_n$ involves the previous recurrence coefficients, as well as a factor of the form

$$\omega_{n+1}^2 \|r_n^{\mathrm{BCG}}\|^2 \tag{7.6}$$

Hence, the standard choice $\omega_{n+1} = 1$ would require the knowledge of $\|r_n^{\mathrm{BCG}}\|$. However, in the QMRS algorithm, only the squared residual vector $r_{2n}^{\mathrm{CGS}}$ is available, instead of $r_n^{\mathrm{BCG}}$. Hence the choice $\omega_{n+1} = 1$ is not feasible, and instead we set

$$\omega_{n+1} := \frac{\sqrt{\|r_n^{\mathrm{CGS}}\| \cdot \|r_0\|}}{\|r_n^{\mathrm{BCG}}\|}. \tag{7.7}$$

Note that, with (7.7), the factor (7.6) reduces to

$$\|r_n^{\mathrm{CGS}}\| \cdot \|r_0\|,$$

and consequently, the coefficients $\mu_n$ and $\nu_n$ corresponding to the weights (7.7) can be updated in the course of the QMRS algorithm. Note that, with (4.2)–(4.3), we can express (7.7) in the form

$$\omega_{n+1} := \frac{\sqrt{\|(\varphi_n^{\mathrm{BCG}}(A))^2 r_0\| \cdot \|r_0\|}}{\|\varphi_n^{\mathrm{BCG}}(A) r_0\|},$$

which suggests that $\omega_{n+1} \approx 1$.

Finally, we state the resulting implementation of QMRS without look-ahead and with weights (7.7). As in § 4.3, we formulate the algorithm in terms of the preconditioned system linear system (4.16).

ALGORITHM 7.1. (QMRS WITHOUT LOOK-AHEAD; WITH WEIGHTS (7.7))

*1. Start:*

*(a) Choose* $x_0 \in \mathbb{C}^N$ *and set* $z_0 = x'_0 = M_2 x_0$.

*(b) Set* $s_0 = t_0 = p_0 = u_0 = w_0 = M_1^{-1} b - A' x'_0$, $v_0 = A' p_0$, $f_0 = 0$;

$$\xi_0 = \|w_0\|^2,\ \eta_0 = 0,\ \mu_0 = 0,\ \nu_0 = 1.$$

*(c) Choose* $\tilde{r}_0$ *such that* $\rho_0 = \tilde{r}_0^H w_0 \neq 0$.

*2. For* $n = 1, 2, \cdots$, *do:*

*(a) Set* $\sigma_{n-1} = \tilde{r}_0^H v_{n-1}$, $\alpha_{n-1} = \rho_{n-1}/\sigma_{n-1}$;

$g_n = u_{n-1} - \alpha_{n-1} v_{n-1}$,

$f_n = \mu_{n-1} f_{n-1} + \nu_{n-1} z_{n-1} + \alpha_{n-1} s_{n-1}$;

$z_n = z_{n-1} + \alpha_{n-1}(u_{n-1} + g_n)$, $w_n = w_{n-1} - \alpha_{n-1} A'(u_{n-1} + g_n)$.

*(b) Set* $\eta_n = \|w_n\| \cdot \|r_0\|/\xi_{n-1}$, $\nu_n = 1/(1+\eta_n)$,

$\mu_n = \eta_n \nu_n$, $\xi_n = \xi_{n-1} \mu_n$;

$x'_{2n} = \mu_n^2 x'_{2n-2} + 2\mu_n \nu_n f_n + \nu_n^2 z_n$.

*If* $x'_{2n}$ *has converged, then set* $x_{2n} = M_2^{-1} x'_{2n}$ *and stop*.

*(c) Set* $\rho_n = \tilde{r}_0^H w_n$, $\beta_n = \rho_n/\rho_{n-1}$;

$u_n = w_n + \beta_n g_n$, $p_n = u_n + \beta_n(g_n + \beta_n p_{n-1})$;

$v_n = A' p_n$, $y_n = t_{n-1} - \alpha_{n-1} A' s_{n-1}$;

$t_n = \nu_n w_n + \mu_n y_n$, $s_n = t_n + \beta_n(\nu_n g_n + \mu_n s_{n-1})$.

We remark that, for simplicity, we have set $w_n := M_1^{-1} r_{2n}^{\mathrm{CGS}}$ and $z_n := M_2 x_{2n}^{\mathrm{CGS}}$ in Algorithm 7.1.

**8. Numerical examples.** In this section, we present numerical results for a few examples. The figures always show the relative residual norms $\|r_n\|/\|r_0\|$ (for QMR Algorithm 3.1 and standard QMR), respectively $\|r_{2n}\|/\|r_0\|$ (for TFQMR, QMRS, and CGS), plotted versus the iteration index $n$.

**8.1. Example 1.** Our first example is a real linear system (3.6) with a nonsingular symmetric indefinite coefficient matrix $T$ of the form (3.7). The order of $T$ is $N = 1000$, and the upper left block $H$ in (3.7) is a positive definite $600 \times 600$ matrix. The matrix $T$ is highly indefinite with 600 positive and 400 negative eigenvalues. For this system we obtained a symmetric indefinite preconditioner by computing a Bunch-Parlett factorization [2] of the form

$$M = LDL^T \tag{8.1}$$

of an approximation $M$ to $T$. Here $L$ is a unit lower triangular matrix and $D$ is a nonsingular block diagonal matrix with $1 \times 1$ and $2 \times 2$ blocks.

We ran the ideal transpose-free QMR Algorithm 3.1 with $J = I_N$ and preconditioner $M$, and we compared it to MINRES [33]. Recall that MINRES can only be used with a symmetric positive definite preconditioner $\hat{M}$. Following a proposal made in [23], we used the matrix $\hat{M} = L\hat{D}L^T$ that is obtained from (8.1) by constructing a block diagonal matrix $\hat{D}$ whose eigenvalues are just the absolute values of those of the indefinite matrix $D$. The resulting convergence curves for QMR (solid line) and MINRES (dotted line) are shown in Fig. 8.1.

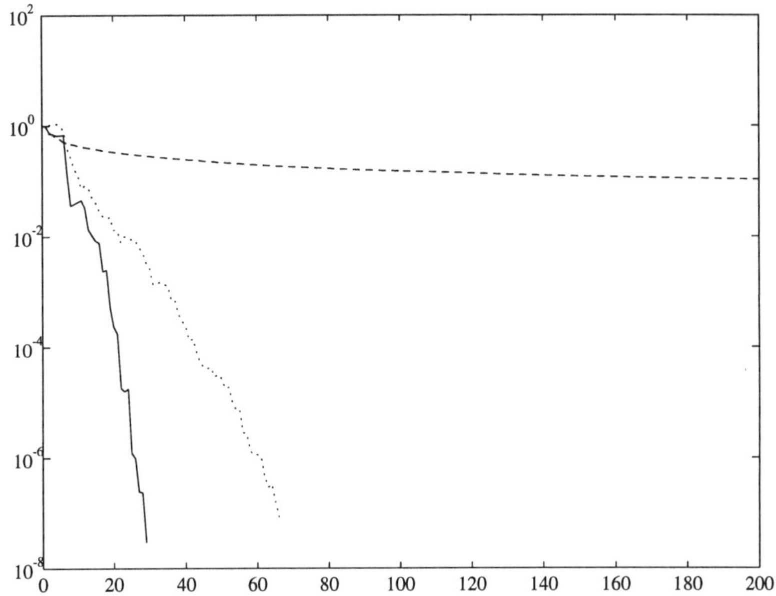


FIG. 8.1. *Convergence curves for Example 1.*

In this example, QMR converges more than twice as fast as MINRES. Note that both algorithms have the same work and storage requirements per iteration.

If no preconditioning is used, then the QMR Algorithm 3.1 and MINRES are mathematically equivalent. In Fig. 8.1, we also plotted the residual norm for a run of Algorithm 3.1 without look-ahead (dashed line). Note that the method does not converge in a reasonable number of iterations. We remark that all Krylov subspace iterations typically converge very slowly for highly indefinite coefficient matrices.

The next two examples are linear systems (2.1) with nonsingular nonsymmetric coefficient matrices. In both cases, we ran the TFQMR Algorithm 4.1, the QMRS Algorithm 7.1, and CGS.

**8.2. Example 2.** This example is the SHERMAN5 matrix from the Harwell-Boeing set of sparse test matrices [6]. This is a matrix of order $N = 3312$, and it has 20793 nonzero elements. This example was run without preconditioning. The convergence curves for QMRS (solid line), TFQMR (dashed line), and CGS (dotted line) are shown in Fig. 8.2. Note that CGS does diverge, while QMRS and TFQMR converge.

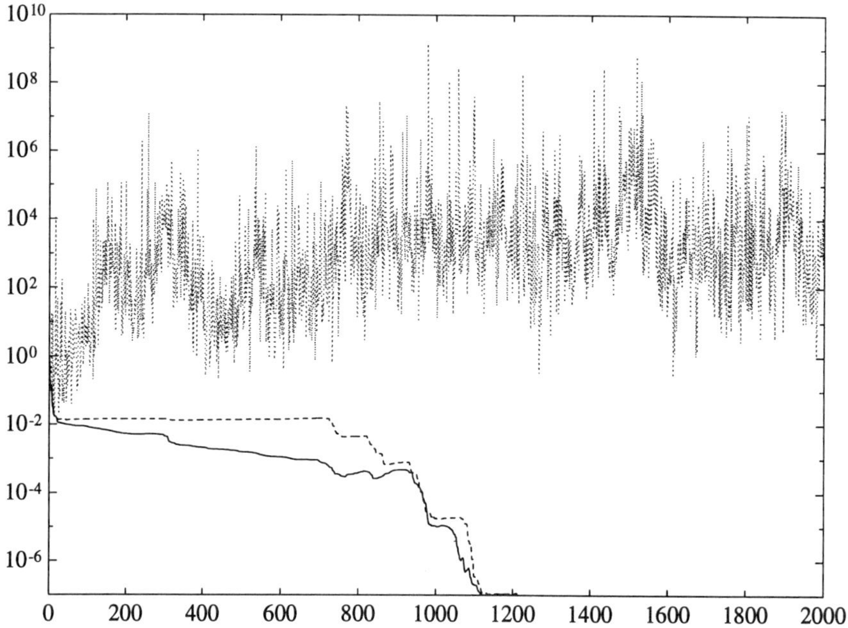


FIG. 8.2. *Convergence curves for Example 2.*

**8.3. Example 3.** This example is a linear system arising from the partial differential equation

$$-\Delta u + 40\left(x\frac{\partial u}{\partial x} + y\frac{\partial u}{\partial y} + z\frac{\partial u}{\partial z}\right) - 240u = f \quad \text{on} \quad (0,1)\times(0,1)\times(0,1).$$

We discretize this equation using centered differences on a uniform $25 \times 25 \times 25$ grid with mesh size $h = 1/26$. The resulting linear system has a sparse coefficient matrix $A$ of order $N = 15625$ with 105625 nonzero elements. For this example, we also ran the standard QMR algorithm. All algorithms were run with a variant [18] of an ILU preconditioner due to Saad [35]. The convergence curves for QMRS (solid line), TFQMR (dashed line), CGS (dotted line), and standard QMR (dash-dotted line) are shown in Fig. 8.3.

**9. Concluding remarks and open problems.** The standard QMR method for solving general non-Hermitian linear systems requires matrix-vector products with both the coefficient matrix and its transpose. We have discussed various QMR-type approaches that do not involve the transpose.

We have shown that, in principle, it is always possible to eliminate the need for the transpose in the standard QMR method by choosing special starting vectors. However, the resulting ideal transpose-free QMR method is practical only for special cases, such as complex symmetric, symmetric indefinite, or Toeplitz systems. We have then described two transpose-free QMR schemes for general non-Hermitian matrices, the TFQMR and QMRS methods, and we presented implementations of both approaches for the case that no look-ahead steps are taken. We stress that standard QMR, TFQMR, and QMRS are not mathematically equivalent. In particular, the sets of iterates generated by the three algorithms are different in general. Our numerical experiments with TFQMR and QMRS show that both algorithms usually display a similar convergence behavior. However, since QMRS requires three matrix-vector multiplications with the coefficient matrix per iteration, compared to only two multiplications per TFQMR iteration, the TFQMR algorithm appears to be preferable to QMRS.

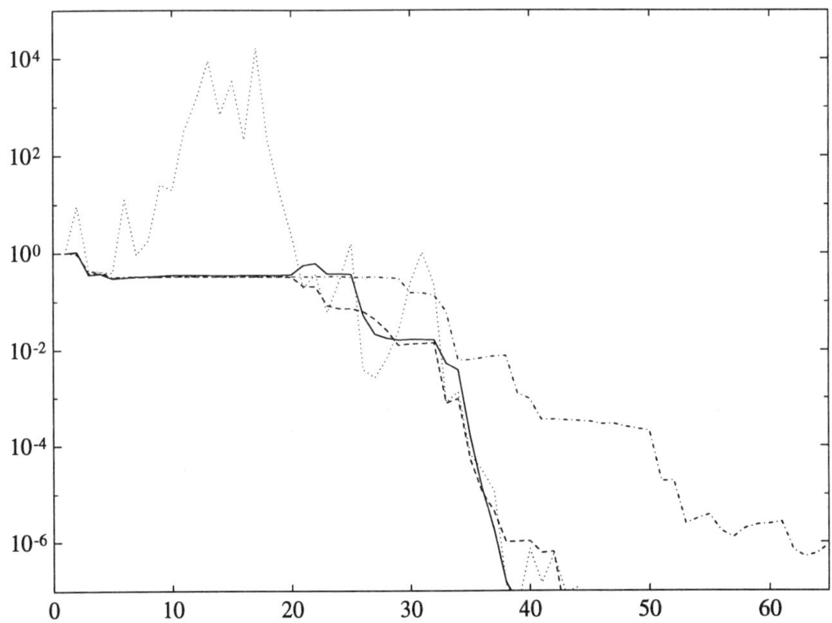


FIG. 8.3. *Convergence curves for Example 3.*

We remark that Chan, de Pillis, and van der Vorst [4] have proposed yet another transpose-free QMR algorithm. Their scheme is mathematically equivalent to the standard QMR method without look-ahead. The use of the transpose is avoided by generating the Lanczos tridiagonal matrix $T_n$ in (2.8) by means of a squared transpose-free Lanczos process. Once $T_n$ is computed, the standard Lanczos vectors $\{v_j\}_{j=1}^{n+1}$ are obtained by running the recurrences summarized in the first identity in (2.8). Finally, the QMR iterates are then computed as in the standard QMR algorithm, based on their definition (2.13)–(2.14). Like QMRS, the resulting transpose-free QMR scheme also requires three matrix-vector multiplications per step, and thus it is not competitive with TFQMR. Furthermore, the underlying squared Lanczos process appears to be less robust than the standard Lanczos algorithm, and there are cases where this transpose-free QMR algorithm converges much slower than standard QMR or where it even diverges.

We have defined the TFQMR and QMRS methods in the general setting of the look-ahead Lanczos process. However, actual implementations were only given for TFQMR and QMRS without look-ahead, and it cannot be excluded that exact breakdowns or near-breakdowns occur in both algorithms. We remark that, in exact arithmetic, the conditions for an exact breakdown in CGS, TFQMR, or QMRS are equivalent. To obtain robust implementations of these three methods, it is crucial to incorporate look-ahead steps that allow to skip over those iterations in which exact or near-breakdowns would occur. It is an open question, whether such look-ahead versions of CGS and TFQMR can be implemented with short vector updates and with only two matrix-vector multiplications per iteration. We remark that the problem of avoiding exact breakdowns in the CGS algorithm was studied in [1].

It was observed that Krylov subspace iterations based on the Lanczos process are typically numerically more robust than their transpose-free counterparts. While it is certainly attractive to eliminate the need for matrix-vector products with the transpose, it appears that the use of the transpose enhances the numerical stability of Lanczos-based Krylov subspace iterations. This phenomenon is not at all understood, and we feel that more research is needed to clarify the role of the transpose.

**10. Acknowledgment.** The author would like to thank Tedd Szeto for his help with the numerical examples in § 8.2 and § 8.3.

## REFERENCES


[1] C. BREZINSKI AND H. SADOK, *Avoiding breakdown in the CGS algorithm*, Numer. Algorithms, 1 (1991) pp. 199–206.

[2] J. R. BUNCH AND B. N. PARLETT, DIRECT METHODS FOR SOLVING SYMMETRIC INDEFINITE SYSTEMS OF LINEAR EQUATIONS, SIAM J. Numer. Anal., 8 (1971), pp. 639–655.

[3] R. H. CHAN AND G. STRANG, *Toeplitz equations by conjugate gradients with circulant preconditioner*, SIAM J. Sci. Stat. Comput., (1989), pp. 104–119.

[4] T. F. CHAN, L. DE PILLIS, AND H. A. VAN DER VORST, *A transpose-free squared Lanczos algorithm and application to solving nonsymmetric linear systems*, Tech. Report CAM 91-17, Department of Mathematics, University of California, Los Angeles, California, October 1991.

[5] P. CONCUS, G. H. GOLUB, AND D. P. O'LEARY, *A generalized conjugate gradient method for the numerical solution of elliptic partial differential equations*, in *Sparse Matrix Computations*, J. R. Bunch and D. J. Rose, eds., Academic Press, New York, 1976, pp. 309–332.

[6] I. S. DUFF, R. G. GRIMES, AND J. G. LEWIS, *Sparse matrix test problems* , ACM Trans. Math. Softw., 15 (1989), pp. 1–14.

[7] S. C. EISENSTAT, *Efficient implementation of a class of preconditioned conjugate gradient methods*, SIAM J. Sci. Stat. Comput., 2 (1981), pp. 1–4

[8] V. FABER AND T. MANTEUFFEL, *Necessary and sufficient conditions for the existence of a conjugate gradient method*, SIAM J. Numer. Anal., 21 (1984) pp. 352–362

[9] R. FLETCHER, *Conjugate gradient methods for indefinite systems* in *Proc. Dundee Conference on Numerical Analysis*, 1975, Lecture Notes in Mathematics 506, G. A. Watson, ed., Springer, Berlin, 1976 pp. 73–89.

[10] R. W. FREUND, *Pseudo-Ritz values for indefinite Hermitian matrices*, Tech. Report 89.33, RIACS, NASA Ames Research Center, Moffett Field, California, August 1989.

[11] ———, *Conjugate gradient-type methods for linear systems with complex symmetric coefficient matrices*, SIAM J. Sci. Stat. Comput., 13 (1992), pp. 425–448.

[12] ———, *A transpose-free quasi-minimal residual algorithm for non-Hermitian linear systems*, SIAM J. Sci. Stat. Comput. (to appear).

[13] ———, *Quasi-kernel polynomials and their use in non-Hermitian matrix iterations,* J. Comput. Appl. Math. (to appear).

[14] ———, *Quasi-kernel polynomials and convergence results for quasi-minimal residual iterations,* in *Numerical Methods of Approximation Theory,* D. Braess and L. L. Schumaker, eds., Birkäuser, Basel, 1992 (to appear).

[15] R. W. FREUND, G. H. GOLUB, AND N. M. NACHTIGAL, *Iterative solution of linear systems*, Acta Numerica, 1 (1992) pp. 57–100.

[16] R. W. FREUND, M. H. GUTKNECHT, AND N. M. NACHTIGAL, *An implementation of the look-ahead Lanczos algorithm for non-Hermitian matrices*, SIAM J. Sci. Stat. Comput. (to appear).

[17] R. W. FREUND AND M. HOCHBRUCK, *On the use of two QMR algorithms for solving singular systems and applications in Markov chain modeling*, Tech. Report 91.25, RIACS, NASA Ames Research Center, Moffett Field, California, December 1991.

[18] R. W. FREUND AND N. M. NACHTIGAL, *QMR: a quasi-minimal residual method for nonHermitian linear systems*, Numer. Math., 60 (1991) pp. 315–339.

[19] ———, *An implementation of the QMR method based on coupled two-term recurrences*, Numerical Analysis Manuscript 92–06, AT&T Bell Laboratories, Murray Hill, New Jersey, July 1992.

[20] R. W. FREUND AND T. SZETO, *A quasi-minimal residual squared algorithm for non-Hermitian linear systems*, Tech. Report 91.26, RIACS, NASA Ames Research Center, Moffett Field, California, December 1991.

[21] R. W. FREUND AND H. ZHA, *A look-ahead algorithm for the solution of general Hankel systems*, Numer. Math. (to appear).

[22] ———, *Simplifications of the nonsymmetric Lanczos process and a new algorithm for solving symmetric indefinite linear systems*, Tech. Report, RIACS, NASA Ames Research Center, Moffett Field, California, 1992.
[23] P. E. Gill, W. Murray, D. B. Ponceleón, and M. A. Saunders, *Preconditioners for indefinite systems arising in optimization*, SIAM J. Matrix Anal. Appl., 13 (1992) pp. 292–311.
[24] G. H. Golub and C. F. Van Loan, *Matrix Computations*, The Johns Hopkins University Press, Baltimore, 2nd ed., 1989.
[25] M. D. Gunzburger, *Finite Element Methods for Viscous Incompressible Flows*, Academic Press, San Diego, 1989.
[26] M. H. Gutknecht , *A completed theory of the unsymmetric Lanczos process and related algorithms, Part I*, SIAM J. Matrix Anal. Appl., 13 (1992) pp. 594–639.
[27] M. R. Hestenes and E. L. Stiefel, *Methods of conjugate gradients for solving linear systems*, J. Res. Nat. Bur. Standards, 49 (1952) pp. 409–436.
[28] R. A. Horn and C. R. Johnson, *Matrix Analysis*, Cambridge University Press, Cambridge, 1985.
[29] W. D. Joubert and D. M. Young, *Necessary and sufficient conditions for the simplification of generalized conjugate-gradient algorithms*, Linear Algebra Appl., 88/89 (1987) pp. 449–485.
[30] J. A. Meijerink and H. A. van der Vorst, *An iterative solution method for linear systems of which the coefficient matrix is a symmetric $M$-matrix*, Math. Comp., 31 (1977) pp. 148–162.
[31] C. Lanczos, *An iteration method for the solution of the eigenvalue problem of linear differential and integral operators*, J. Res. Nat. Bur. Standards, 45 1950, 255–282.
[32] ———, *Solution of systems of linear equations by minimized iterations*, J. Res. Nat. Bur. Standards, 49 (1952) pp. 33–53.
[33] C. C. Paige and M. A. Saunders , *Solution of sparse indefinite systems of linear equations*, SIAM J. Numer. Anal., 12 (1975) pp. 617–629.
[34] B. N. Parlett, D. R. Taylor, and Z. A. Liu, *A look-ahead Lanczos algorithm for unsymmetric matrices*, Math. Comp., 44 (1985) pp. 105–124.
[35] Y. Saad, *ILUT: a dual threshold incomplete LU factorization*, Tech. Report UMSI 92/38, University of Minnesota Supercomputer Institute, Minneapolis, Minnesota, March 1992.
[36] Y. Saad and M. H. Schultz, *GMRES: a generalized minimal residual algorithm for solving nonsymmetric linear systems*, SIAM J. Sci. Stat. Comput., 7 (1986) pp. 856–869.
[37] P. Sonneveld, *CGS, a fast Lanczos-type solver for nonsymmetric linear systems*, SIAM J. Sci. Stat. Comput., 10 (1989) pp. 36–52 .
[38] E. L. Stiefel, *Kernel polynomials in linear algebra and their numerical applications*, U.S. National Bureau of Standards, Applied Mathematics Series 49 (1958) pp. 1–22.
[39] V. V. Voevodin , *The question of non-self-adjoint extension of the conjugate gradients method is closed*, USSR Comput. Math. and Math. Phys., 23 (1983) pp. 143–144.
[40] M. H. Wright, *Interior methods for constrained optimization*, Acta Numerica, 1 (1992) pp. 341–407.

# MATRICES THAT GENERATE THE SAME KRYLOV RESIDUAL SPACES

ANNE GREENBAUM* AND ZDENEK STRAKOS†

**Abstract.** Given an $n$ by $n$ nonsingular matrix $A$ and an $n$-vector $v$, we consider the spaces of the form $AK_k(A,v)$, $k = 1,...,n$, where $K_k(A,v)$ is the $k^{th}$ Krylov space, equal to $span\{v, Av, ..., A^{k-1}v\}$. We characterize the set of matrices $B$ that, with the given vector $v$, generate the same spaces; i.e., those matrices $B$ for which $BK_k(B,v) = AK_k(A,v)$, for all $k = 1,...,n$. It is shown that any such sequence of spaces can be generated by a unitary matrix. If zero is outside the field of values of $A$, then there is a Hermitian positive definite matrix that generates the same spaces, and, moreover, if $A$ is close to Hermitian then there is a nearby Hermitian matrix that generates the same spaces. It is also shown that any such sequence of spaces can be generated by a matrix having any desired eigenvalues.

Implications about the convergence rate of the GMRES method are discussed. A new proof is given that if zero is outside the field of values of $A$, then convergence of the GMRES algorithm is strictly monotonic. It is shown that if $A$ is close to Hermitian, then the eigenvalues of $A$ essentially determine the behavior of the GMRES iteration but that, in general, eigenvalue information alone is never sufficient to ensure rapid convergence of the GMRES algorithm.

## 1. Introduction.

Most iterative methods for solving a linear system

$$Ax = b \tag{1.1}$$

start with an initial guess $x^0$ for the solution and, at each step $k$, generate an approximate solution $x^k$ from the linear variety

$$x^0 + span\{r^0, Ar^0, ..., A^{k-1}r^0\}, \tag{1.2}$$

where $r^0 \equiv b - Ax^0$ is the initial residual. Several of these methods – e.g., (full) GMRES[1] [16], ORTHOMIN [21], and ORTHODIR [22] – choose the approximation $x^k$ to minimize the Euclidean norm of the residual $r^k \equiv b - Ax^k$. A question of considerable interest is how accurate an approximation $x^k$ can be obtained from this linear variety.

* Courant Institute of Mathematical Sciences, 251 Mercer St., New York, NY 10012. This work was supported in part by DOE contract DEFG0288ER25053. Part of this work was performed while visiting IBM T.J. Watson Research Center and part was performed while visiting the Institute for Mathematics and its Applications, University of Minnesota.

† Institute of Computer and Information Sciences, Czech Academy of Sciences, Pod vod vezi 2, 18207 Praha 8, Czechoslovakia. This work was performed while visiting the Institute for Mathematics and its Applications, University of Minnesota.

[1] Throughout this paper we will refer to the GMRES algorithm, but since we always assume exact arithmetic, our statements will apply equally well to any mathematically equivalent formulation.

The residual vector $r^k$ is the vector of smallest 2-norm in the linear variety

$$r^0 + span\{Ar^0, ..., A^k r^0\}. \tag{1.3}$$

It follows that $r^k$ can be written in the form

$$r^k = P_k(A) r^0, \tag{1.4}$$

where $P_k$ is the $k^{th}$ degree polynomial with value one at the origin that minimizes $\|p_k(A)r^0\|$ over the set $\mathcal{P}_k$ of all $k^{th}$ degree polynomials $p_k$ with $p_k(0) = 1$.

A simple bound on the size of the residual can be derived as follows. If we assume that the matrix $A$ has a complete set of eigenvectors $Z$, so that $A$ can be written in the form $A = Z\Lambda Z^{-1}$, where $\Lambda = diag(\lambda_1, ..., \lambda_n)$ is the diagonal matrix of eigenvalues, then $r^k$ can be written in the form

$$r^k = Z P_k(\Lambda) Z^{-1} r^0. \tag{1.5}$$

For any other $k^{th}$ degree polynomial $p_k \epsilon \mathcal{P}_k$, we have

$$\|r^k\| \leq \|Z p_k(\Lambda) Z^{-1} r^0\| \leq \kappa(Z) \max_{i=1,...,n} |p_k(\lambda_i)| \, \|r^0\|, \tag{1.6}$$

where $\kappa(Z) = \|Z\| \, \|Z^{-1}\|$ is the condition number of $Z$ and $\|\cdot\|$ here, and elsewhere, denotes the 2-norm for vectors and the corresponding spectral norm for matrices. The bound in (1.6) is minimized by taking $p_k$ to be the $k^{th}$ degree minimax polynomial on the discrete set of eigenvalues $\{\lambda_1, ..., \lambda_n\}$, and (1.6) can be written in the form

$$\frac{\|r^k\|}{\|r^0\|} \leq \kappa(Z) \min_{p_k \epsilon \mathcal{P}_k} \max_{i=1,...,n} |p_k(\lambda_i)|. \tag{1.7}$$

The columns of the matrix $Z$ can be normalized arbitrarily in order to make $\kappa(Z)$ as small as possible. Throughout this paper, however, we will assume that the columns of $Z$ have norm 1. This scaling usually gives a near optimal value for $\kappa(Z)$.

If the matrix $A$ is *normal*, then $\kappa(Z)$ is one, and it was recently shown in several different ways [8,9,11] that this bound is *sharp*; i.e., that for each $k$, there is an initial vector $r^0$ (depending on $k$) for which equality holds in (1.7). In many cases of interest, however, the matrix $A$ is not normal and the factor $\kappa(Z)$ in (1.7) may be quite large. (See, for instance, [15] for some interesting physical examples.) In such cases, the bound (1.7) may be a large overestimate of the actual residual.

To deal with such problems, Trefethen has introduced bounds based on the *$\epsilon$-pseudo-eigenvalues* of the matrix [18]. There are two equivalent definitions of the $\epsilon$-pseudo-spectrum of a matrix $A$:

**Definition 1.** The $\epsilon$-pseudo-spectrum of $A$, denoted $\Lambda_\epsilon(A)$, is the set of points $z \in C$ such that $\|(zI - A)^{-1}\| \geq \epsilon^{-1}$.

**Definition 2.** The $\epsilon$-pseudo-spectrum of $A$ is the set of points $z \in C$ which are eigenvalues of some matrix $A + E$ with $\|E\| \leq \epsilon$.

For any analytic function $f$, it is known [2, p. 560] that $f(A)$ can be written in the form

$$f(A) = \frac{1}{2\pi i} \int_\Gamma (zI - A)^{-1} f(z) dz, \tag{1.8}$$

where $\Gamma$ is any simple closed curve or union of simple closed curves containing the spectrum of $A$. Taking norms on both sides and replacing the norm of the integral by the length $\mathcal{L}(\Gamma)$ of the curve times the maximum norm of the integrand factors gives

$$\|f(A)\| \leq \frac{1}{2\pi} \mathcal{L}(\Gamma) \max_{z \in \Gamma} \|(zI - A)^{-1}\| \max_{z \in \Gamma} |f(z)|. \tag{1.9}$$

Note that the inequality (1.9) will be close to an equality if and only if the integrand in (1.8) is nearly constant in norm over the curve $\Gamma$ and cancellation does not cause the norm of the integral to be much smaller than the length of the curve times the norm of the integrand.

A reasonable set of curves to consider, then, are the curves $\Gamma_\epsilon$ on which the resolvent $(zI - A)^{-1}$ has constant norm $1/\epsilon$; i.e., the boundaries of the $\epsilon$-pseudo-spectra of $A$. In this case, the bound (1.9) becomes

$$\|f(A)\| \leq \frac{\mathcal{L}(\Gamma_\epsilon)}{2\pi\epsilon} \max_{z \in \Gamma_\epsilon} |f(z)|. \tag{1.10}$$

Taking the function $f$ to be the polynomial $p_k \in \mathcal{P}_k$ for which this bound is minimal gives the following bound on the residual in (1.4):

$$\frac{\|r^k\|}{\|r^0\|} \leq \frac{\mathcal{L}(\Gamma_\epsilon)}{2\pi\epsilon} \min_{p_k \in \mathcal{P}_k} \max_{z \in \Gamma_\epsilon} |p_k(z)|. \tag{1.11}$$

While, for properly chosen values of $\epsilon$, the bound (1.11) sometimes gives a much better estimate than (1.7) of the actual size of the residual, there are other cases in which even the bound (1.11) is a large overestimate, for any value of $\epsilon$. Consider, for example, a matrix $A$ of the form $Z\Lambda Z^{-1}$, where

$$Z = \begin{pmatrix} 1 & \sqrt{1-\delta} & 0 & \dots & 0 \\ 0 & \sqrt{\delta} & 0 & \dots & 0 \\ 0 & 0 & 1 & \dots & 0 \\ \cdot & \cdot & \cdot & \cdot & \cdot \\ 0 & 0 & 0 & \dots & 1 \end{pmatrix}, \quad \delta << 1, \quad \Lambda = \begin{pmatrix} 20 & & & & \\ & 10 & & & \\ & & 5 & & \\ & & & \cdot & \\ & & & & 1 \end{pmatrix}. \tag{1.12}$$

$A$ has two large, well-separated eigenvalues, 20 and 10, corresponding to an ill-conditioned block in $Z$. The remaining eigenvalues are uniformly distributed in the interval $[1, 5]$. The condition number of $Z$ is approximately $2/\sqrt{\delta}$.

Fig. 1a shows the $\epsilon$-pseudo-spectra of such a matrix $A$, of order $n = 19$ and with $\delta = 10^{-8}$, for $\epsilon = 10^{-5}$ and $10^{-4}$. (Note that the scales on the horizontal and vertical axes in the figure are different. The pseudo-spectral regions are approximately circular.) Note that the factor $\frac{\mathcal{L}(\Gamma_\epsilon)}{2\pi\epsilon}$ in (1.11) is on the order of $2 \cdot 10^4$ for these values of $\epsilon$, and a low degree polynomial $p_k$ could not be small enough on these curves to make the bound (1.11) reasonable; i.e., less than 1 for $k << 19$. For larger values of $\epsilon$ ($\epsilon >\sim 10^{-2}$), the $\epsilon$-pseudo-spectrum contains the origin, and then, because of the maximum modulus theorem, the bound (1.11) cannot be useful since $p_k(0) = 1$.

Yet for this problem the GMRES iteration will behave almost as it would if applied to the diagonal matrix $\Lambda$. It will annihilate the top 2 by 2 block quickly, just as it would annihilate the two large eigenvalues 20 and 10 of $\Lambda$ quickly, and then converge as if it were working with a diagonal matrix whose eigenvalues lie between 1 and 5. Fig. 1b shows the convergence of the GMRES method applied to a linear system with coefficient matrix $A$ (solid line) and to one with coefficient matrix $\Lambda$ (dashed line), with a random initial residual. The + marks in Fig. 1b show a *sharp* upper bound on the residual at each step of the GMRES algorithm applied to $A$. (This is also an upper bound on the residual at each step of the GMRES algorithm applied to $\Lambda$, and it is very similar to the sharp upper bound for this problem.) This was computed using an optimization code (*fminu* in *MATLAB* [13]) to find, for each $k$, an initial residual $r^0$ for which $||r^k||/||r^0||$ was as large as possible. These bounds were checked in the following way. Taking the initial vector $r^0$, with $||r^0|| = 1$, returned by the optimization routine, we computed the GMRES polynomial $P_k(A)$ in (1.4). If $||P_k(A)||$ is equal to $||r^k||$, then this must be a sharp upper bound, since it is attainable and since, for any other initial residual $\hat{r}^0$, we have $||\hat{r}^k|| \leq ||P_k(A)|| \; ||\hat{r}^0||$.

Although it often required carefully chosen initial guesses to enable the optimization code to find the worst initial residual, the bounds plotted in Fig. 1b were all proven correct in this way. The bound (1.11) based on the $\epsilon$-pseudo-spectra and the bound (1.7) based on the condition number of the eigenvector matrix are both *much* larger than the sharp bound pictured in Fig. 1b.

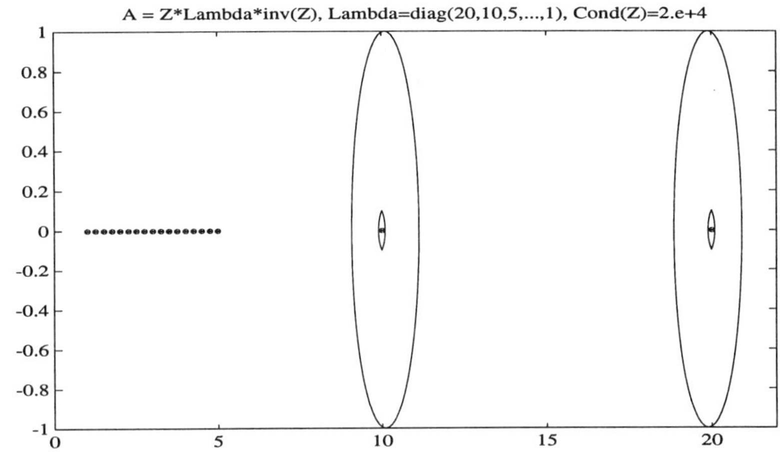


Fig. 1a. 1.e-5 and 1.e-4 Pseudo-spectra

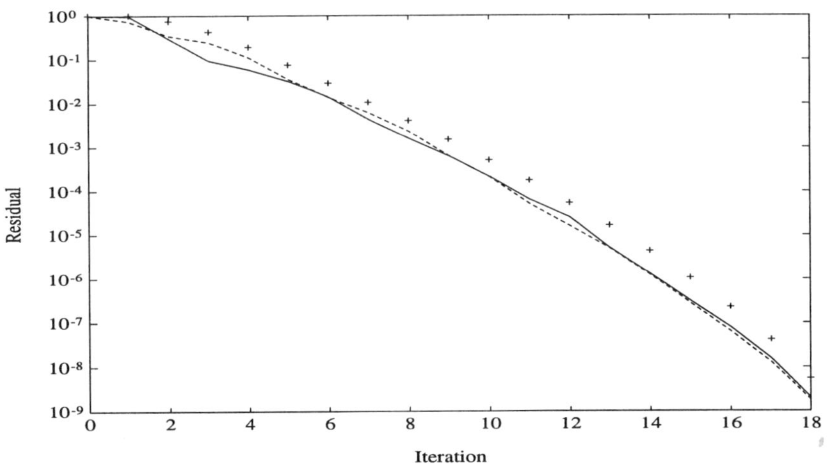


Fig. 1b. GMRES(A) (solid), GMRES(Lambda)(dashed), Sharp Bound(+)

It should also be noted that if $A$ is reduced to Schur form, then the Frobenius norm of the strict upper triangle of the Schur form is also quite large – in this case about $10^5$. Thus, the matrix $A$ is *far* from normal in all of the usual senses, and yet the GMRES iteration converges very well.

The problem of deriving a sharp bound on the convergence rate of the GMRES method, in terms of eigenvalues or pseudo-eigenvalues or other simple properties of the matrix, appears to be a difficult one. In this paper, we take a different approach to this problem. While it does not yet achieve the ultimate goal of giving a sharp or nearly sharp bound for all cases, it does yield certain insights into the behavior of the GMRES method when applied to linear systems with non-normal coefficient matrices. We relate the behavior of the method applied to a non-normal matrix to its behavior when applied to certain normal matrices. Since the convergence rate of the method applied to normal matrices is determined by the eigenvalues of the matrix, if the eigenvalues of the normal matrix could be related to some properties of the original matrix, then (1.7) would give a bound on the convergence rate, in terms of these properties.

Unfortunately, it is only in special cases that we are able to relate the eigenvalues of the normal matrix to meaningful properties of the original matrix. It is shown that any behavior that can be seen with the GMRES method can be seen with the method applied to a unitary matrix. While certain properties of the original matrix – e.g., positive definiteness – appear to guarantee large gaps in the spectrum of this unitary matrix, we have not been able to prove this. It is shown that if zero is outside the field of values of the original matrix $A$, then the GMRES method behaves just as it would for a certain Hermitian positive definite matrix. If $A$ is close to Hermitian, then the GMRES method behaves just as it would for a Hermitian matrix whose eigenvalues are close to those of $A$. In this case, then, the eigenvalues of $A$ essentially determine the behavior of the GMRES algorithm. In general, however, eigenvalue information alone cannot be sufficient to ensure fast convergence of the GMRES algorithm. Any behavior that can be seen with the method can be seen with the method applied to a matrix having *any* nonzero eigenvalues.

Throughout this paper, matrices and vectors will be assumed to be complex, unless otherwise stated. A superscript $^*$ will denote the Hermitian transpose. The space spanned by a set of vectors $\{v^1, ..., v^k\}$ will be denoted $[v^1, ..., v^k]$, and the set of vectors of the form $v$ plus a linear combination of $\{v^1, ..., v^k\}$ will be denoted $v + [v^1, ..., v^k]$. The symbol $\|\cdot\|$ will always denote the 2-norm of a vector or matrix and $\|\cdot\|_F$ will denote the Frobenius norm of a matrix. The usual Euclidean inner product will be denoted $<\cdot,\cdot>$.

In section 2 we characterize those matrices $B \equiv B(r^0)$ for which the GMRES method, with initial residual $r^0$, generates the same residual vectors at each step as the GMRES method applied to a linear system with coefficient matrix $A$ and initial residual $r^0$. Letting $K_k(A, r^0)$ denote the Krylov space $[r^0, Ar^0, ..., A^{k-1}r^0]$, we know that the residual $r^k$ has the smallest 2-norm among all vectors of the form

$$r^0 + AK_k(A, r^0). \tag{1.13}$$

If the spaces $BK_k(B, r^0)$ are the same as the spaces $AK_k(A, r^0)$, for all $k = 1, ..., n$, then the GMRES method applied to $B$, with initial residual $r^0$, will generate the same residual vectors at each step as the method applied to $A$, with initial residual $r^0$. For lack of a better name, we refer to these spaces as *Krylov residual spaces*, which should not be confused with the Krylov spaces $K_k(A, r^0)$ and $K_k(B, r^0)$. The Krylov spaces $K_k(A, r^0)$ and $K_k(A+\alpha I, r^0)$, where $\alpha$ is any scalar, are the same for all $k$, but the GMRES method behaves very differently when applied to $A$ and to $A + \alpha I$! Note also that it is not required that individual vectors, $Ar^0, ..., A^k r^0$ and $Br^0, ..., B^k r^0$ be equal, only that the spaces they span be the same; i.e., $Br^0$ must be a scalar multiple of $Ar^0$, $B^2 r^0$ must be a linear combination of $A^2 r^0$ and $Ar^0$, etc. Note further that if $r^k$ is written in the form (1.4), $r^k = P_k(A) r^0 = \hat{P}_k(B) r^0$, then the polynomials $P_k$ and $\hat{P}_k$ will be different. Thus, this approach cannot be used to analyze eigenvalue approximations generated by the algorithm, since these will be very different for $A$ and for $B$. In section 3, we consider normal matrices $B$ for which the Krylov residual spaces, with a given initial vector $r^0$, are the same as those of a given matrix $A$. In section 4, we show that any sequence of Krylov residual spaces can be generated by a matrix having any nonzero eigenvalues. Section 5 gives conclusions and discusses remaining open questions and possible applications of this technique of analysis.

## 2. Matrices that generate the same Krylov residual spaces.

Given an $n$ by $n$ nonsingular matrix $A$ and an $n$-vector $v$, we wish to characterize the matrices $B = B(v)$ for which the spaces

$$AK_k(A, v) \equiv [Av, ..., A^k v] \quad \text{and} \quad BK_k(B, v) \equiv [Bv, ..., B^k v] \tag{2.1}$$

are the same, for all $k = 1, ..., n$. Throughout this paper, we will assume that the vectors $Av, ..., A^n v$ are linearly independent. If they are not, however, or if one wishes to consider conditions in which only, say, the first $m$ of these spaces are the same, then the characterizations given here can be modified accordingly. The class of matrices $B$ for which the first $m$ such spaces are the same will, of course, include the class discussed in this paper.

Let $w^1, ..., w^k$ be an orthonormal basis for $[Av, ..., A^k v]$, and let $W$ be the matrix with orthonormal columns $w^1, ..., w^n$. Then it is well-known that the unitary matrix $W$ satisfies

$$AW = WH, \tag{2.2}$$

where $H$ is an unreduced upper Hessenberg matrix. The following theorem characterizes all matrices $B = B(v)$ that generate the same Krylov residual spaces (2.1) with the vector $v$.

THEOREM 2.1. *Using the above notation, let $B$ be of the form*

$$B = WRHW^*, \tag{2.3}$$

*where $R$ is any nonsingular upper triangular matrix. Then*

$$BK_k(B,v) = AK_k(A,v), \quad k = 1,...,n, \tag{2.4}$$

*and, conversely, any matrix $B$ that satisfies (2.4) is of the form (2.3).*

*Proof.* Suppose the order one Krylov residual spaces are the same; i.e.,

$$Bv = cAv \tag{2.5}$$

for some nonzero scalar $c$. The higher order spaces $BK_k(B,v)$ and $AK_k(A,v)$, $k > 1$, will be the same if $B$ satisfies

$$BW = W\tilde{H} \tag{2.6}$$

for some unreduced upper Hessenberg matrix $\tilde{H}$. To see that this is so, assume that $BK_{k-1}(B,v) = AK_{k-1}(A,v)$. The $(k-1)^{st}$ column of equation (2.6) can be written as

$$B\, w^{k-1} = \sum_{j=1}^{k} w^j\, \tilde{H}_{j,k-1}.$$

It follows that $Bw^{k-1}$, and hence $B^k v$, is a linear combination of $w^1, ..., w^k$, and since the coefficient of $w^k$ is nonzero, $B^k v$ is linearly independent of $[w^1, ..., w^{k-1}] = [Bv, ..., B^{k-1}v]$. Thus the order $k$ spaces are the same and the proof is by induction. Conversely, it is clear that (2.5) and (2.6) are necessary conditions in order for (2.4) to be satisfied.

Taking $B$ to be of the form $B = W\tilde{H}W^*$, then, the condition (2.5) becomes, upon multiplying each side by $W^*$,

$$\tilde{H}W^*v = cW^*Av. \tag{2.7}$$

But $W^*Av$ is just a constant times $e_1$, the first unit vector,

$$W^*Av = \hat{c}e_1,$$

and $W^*v$ satisfies

$$W^*v = W^*A^{-1}Av = W^*(WH^{-1}W^*)Av = \hat{c}H^{-1}e_1.$$

Therefore (2.7) can be written as

$$\tilde{H}H^{-1}e_1 = ce_1. \tag{2.8}$$

Clearly, if $\tilde{H}$ is of the form $RH$, where $R$ is a nonsingular upper triangular matrix, then $\tilde{H}$ is an unreduced upper Hessenberg matrix satisfying (2.8). Conversely, if $\tilde{H}$ is an unreduced upper Hessenberg matrix satisfying (2.8), and if we write $\tilde{H}$ in the form $XH$, then the first column of X must be a scalar multiple of $e_1$. But the requirement that $XH$ be upper Hessenberg then implies that the elements below the diagonal in subsequent columns of $X$ are also zero, and the requirement that $XH$ be unreduced implies that the diagonal elements of $X$ are nonzero. Thus, the only matrices $B$ satisfying (2.4) are of the form $B = W\tilde{H}W^*$, where $\tilde{H} = RH$. □

The characterization of Theorem 1 will prove useful in later sections, where we consider normal or near normal matrices of the form $RH$, for various nonsingular upper triangular matrices $R$.

There are many equivalent ways to characterize the matrices $B$ that generate the same Krylov residual spaces. We give two such characterizations which will prove useful in later sections. The following theorem can be derived from Theorem 1, but it is just as easy to prove it directly, which we do below.

THEOREM 2.2. *Using the above notation, suppose $B$ is of the form*

$$B = W\hat{R}\hat{H}W^* \tag{2.9}$$

*where $\hat{R}$ is any nonsingular upper triangular matrix and*

$$\hat{H} = \begin{pmatrix} 0 & . & . & 0 & 1/<v,w^n> \\ 1 & & & . & -<v,w^1>/<v,w^n> \\ & . & & . & . \\ & & . & . & . \\ 0 & . & . & 1 & -<v,w^{n-1}>/<v,w^n> \end{pmatrix}. \tag{2.10}$$

*Then (2.4) holds, and, conversely, any matrix $B$ that satisfies (2.4) is of the form (2.9, 2.10).*

*Proof.* Let $V_A$ be the matrix with columns $(Av, ..., A^nv)$. The condition (2.4) is equivalent to

$$V_AR_1 = B(v, w^1, ..., w^{n-1}), \tag{2.11}$$

for some nonsingular upper triangular matrix $R_1$. To see that this is so, note that (2.11) implies that $Bv$ is a nonzero multiple of $Av$ (and hence of $w^1$), that $Bw^1$ (and hence $B^2v$) is a linear combination of $A^2v$ and $Av$, with the coefficient of $A^2v$ being nonzero, etc. We can also write $V_A = WR_2$, for some nonsingular upper triangular matrix $R_2$, and

$$B(v, w^1, ..., w^{n-1}) = BW \begin{pmatrix} < v, w^1 > & 1 & 0 & . & 0 \\ < v, w^2 > & 0 & 1 & & . \\ . & . & & . & . \\ . & . & & & 1 \\ < v, w^n > & 0 & . & . & 0 \end{pmatrix} \tag{2.12}$$

Substituting these expressions into (2.11) and solving for $B$ gives

$$B = W\hat{R}\hat{H}W^*,$$

where $\hat{R} = R_2R_1$ and $\hat{H}$ is the inverse of the rightmost matrix in (2.12). It is easy to check that this is the matrix $\hat{H}$ defined in (2.10). Since, by assumption, the vectors $Av, ..., A^n v$ are linearly independent, it is also the case that $v, Av, ..., A^{n-1}v$ are linearly independent and hence that $< v, w^n >$ is nonzero. Thus, the matrix $\hat{H}$ is well-defined and the theorem is proved. □

This theorem will be used later to show that any sequence of Krylov residual spaces can be generated by a matrix having any nonzero eigenvalues.

Another characterization of the class of matrices described in Theorems 1 and 2 is given in the following theorem. Let $\mathcal{L}(X)$ denote the lower triangle (including diagonal) of a matrix $X$, and let $\hat{\mathcal{L}}(X)$ denote the strict lower triangle of $X$. The following theorem characterizes the matrices $X$ for which $HX$ is upper triangular.

THEOREM 2.3. *If $H$ is an $n$ by $n$ nonsingular unreduced upper Hessenberg matrix, then $HX$ is upper triangular if and only if the elements of the lower triangle of $X$ satisfy*

$$X_{i-1,j} = -\frac{1}{H_{i,i-1}} \sum_{k=i}^{n} H_{ik}X_{kj}, \quad i = n, n-1, ..., j+1, \quad j = 1, ..., n, \tag{2.13}$$

*where the bottom row of elements $X_{nj}$, $j = 1, ..., n$ is arbitrary. These equations are satisfied by any matrix $X$ whose lower triangle is of the form*

$$\mathcal{L}(X) = \mathcal{L}(H^{-1})\ D, \qquad D \textit{ diagonal}, \tag{2.14}$$

*and if $\mathcal{L}(H^{-1})$ has no zero columns, then all solutions to (2.13) are of the form (2.14).*

*Proof.* The condition that $HX$ be upper triangular means that the elements of the lower triangle of $X$ must satisfy

(2.15) $$(HX)_{ij} = \sum_{k=i-1}^{n} H_{ik}X_{kj} = 0, \quad j < i.$$

Since $H_{i,i-1}$ is nonzero for all $i$, we can solve these equations for $X_{i-1,j}$, $i = n, n-1, ..., j+1$, in terms of $X_{nj}$, to obtain (2.13). If the elements $X_{nj}$ are taken to be of the form $d_j H_{nj}^{-1}$, for some scalars $d_j$, then, since $H^{-1}D$ also satisfies equations (2.15) when $D = diag(d_1, ..., d_n)$, the lower triangle of $X$ must be equal to the lower triangle of $H^{-1}D$. Moreover, all solutions are of this form unless $H^{-1}$ has a zero element in its last row, and, in this case, the equations (2.15) imply that all elements of that column of $\mathcal{L}(H^{-1})$ must be zero. □

According to Theorem 3, matrices $\tilde{H}$ of the form $RH$, as described in Theorem 1, are essentially those matrices whose inverses have lower triangles of the form $\mathcal{L}(H^{-1})D$, or, more generally, of the form (2.13). The strict upper triangle of the inverse of such a matrix is arbitrary. In a later section we will consider the Hermitian matrix obtained by taking the strict upper triangle to be the Hermitian transpose of the strict lower triangle.

**3. Decompositions of the form $H = RN$.** For a given matrix $A$ and vector $v$, let $W$ and $H$ be defined as in the previous section. In light of Theorem 1, it will be instructive to consider matrices $\tilde{H}$ of the form $RH$, for some upper triangular matrices $R$; for if we then define the matrix $B$ to be $W\tilde{H}W^*$, then the Krylov residual spaces generated by $B$ with initial vector $v$ will be the same as those generated by $A$ with initial vector $v$. Since the GMRES algorithm minimizes the 2-norm of the residual vector over the linear variety consisting of the initial residual plus the Krylov residual space, it follows that if the GMRES method is applied to a linear system $Ax = b$, with initial residual $b - Ax^0 = v$, then it will generate the same residual vectors at each step as the GMRES method applied to a linear system $By = f$, with initial residual $f - By^0 = v$. We will express this by writing

(3.1) $$GMRES(A, v) \equiv GMRES(B, v).$$

If, for each vector $v$, we can find a matrix $B$ of the given form, for which we can analyze the behavior of the GMRES method applied to $B$, then we can also analyze the behavior of the GMRES method applied to $A$. Since the behavior of the GMRES method for normal matrices is well-understood in terms of the eigenvalues of the matrix (cf. equation (1.7)), it is desirable to find an upper triangular matrix $R$ such that $RH$ is normal, or, equivalently, to write $H$ in the form $H = \tilde{R}N$, where $\tilde{R}$ is upper triangular (the inverse of the upper triangular matrix in Theorem 1) and $N$ is normal. Alternatively, using Theorem 3, we will look for normal matrices $N$ such that the lower triangle of $N^{-1}$ is equal to the product of the lower triangle of $H^{-1}$ and a diagonal matrix.

There are many decompositions of this form, a few of which will be described in this section. If such a decomposition could be found for which the eigenvalues of $N$ could be related to eigenvalues or pseudo-eigenvalues or other simple properties of the matrix $A$, then the convergence rate of the GMRES method applied to $A$ could be explained in terms of these properties. In general, we have not been able to find such a decomposition, but in special cases this can be done.

**3.1. Equivalent Unitary Matrices.** Any upper Hessenberg matrix $H$ can be written in the form

$$H = RQ, \tag{3.2}$$

where $R$ is upper triangular and $Q$ is unitary (and also upper Hessenberg). Thus, any behavior that can be seen with the GMRES algorithm applied to any matrix can be seen with the GMRES algorithm applied to a unitary matrix! Which unitary matrix will depend on the initial residual as well as the matrix, but for any matrix $A$ and any initial residual $r^0$, if $W$ is the matrix whose columns are the orthonormal basis vectors and $H$ the upper Hessenberg matrix generated after $n$ steps of $GMRES(A, r^0)$, and if $H = RQ$ is the RQ-decomposition of $H$, then

$$GMRES(A, r^0) \equiv GMRES(WQW^*, r^0). \tag{3.3}$$

We have not been able to establish any interesting relationships between the eigenvalues of the unitary matrix $B = WQW^*$ and special properties of $A$, although it appears that such relationships exist, at least in certain cases. Fig. 2a shows the eigenvalues of the unitary matrix $B$ obtained from a GMRES computation with a random 19 by 19 matrix $A$ and a random initial residual. Note that the eigenvalues are fairly uniformly distributed around the unit circle. The solid line in Fig. 2b shows the convergence of the GMRES algorithm, applied to either $A$ or $B$, with this same initial residual, while the dotted line in the figure shows the convergence of the GMRES algorithm applied to $B$ with a different random initial residual, uncorrelated with $B$. Note that the GMRES algorithm behaves similarly, when applied to either $A$ or $B$, with an arbitrary initial vector, indicating that the *typical* behavior of the algorithm applied to $B$ is similar to that of the algorithm applied to $A$. In both cases, we see slow convergence, which is characteristic of the GMRES algorithm applied to a random matrix or to a unitary matrix with eigenvalues all around the unit circle.

Also shown in Fig. 2b are sharp upper bounds for the residual at each step of the GMRES algorithm applied to $A$ (+) and to $B$ (x). These

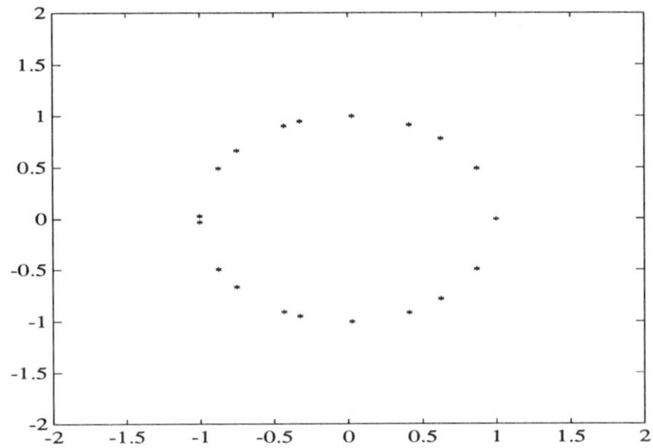

Fig. 2a. Eigenvalues of B. A=rand(19,19)

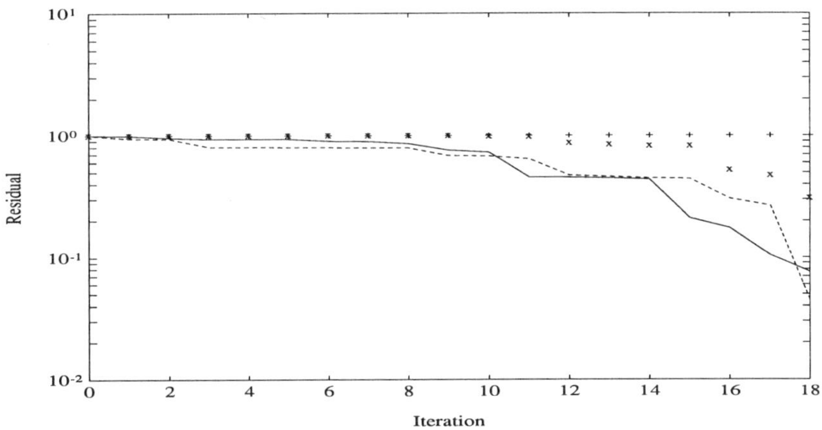


Fig. 2b. GMRMES(A or B,r0)(-) and GMRES(B,random)(–)

were computed numerically and then proven correct, as described earlier. (The initial vector $r^0$ for which $||r^k||/||r^0||$ is maximal was computed using an optimization code or other technique, and then it was determined that $||r^k|| = ||P_k(A)|| \cdot ||r^0||$, where $P_k$ is the GMRES polynomial.) It is interesting to note that the computed sharp bound for $A$ is 1.0000 for steps 1 through 18 – indicating that there is an initial residual for which the GMRES algorithm applied to $A$ makes no progress at all until step $n = 19$! This is frequently observed to be the case for random matrices. Had we used this initial residual for our first experiment, the equivalent unitary matrix $WQW^*$ would have had its eigenvalues uniformly distributed around the unit circle. This can be seen from the characterization of Theorem 2. In this case, $< r^0, w^j >$ is zero for $j = 1, ..., n-1$ and $| < r^0, w^n > | = ||r^0||$, which we can assume without loss of generality is 1. Then the matrix $\hat{H}$ defined in Theorem 2 is a unitary shift matrix, and since the $RQ$ decomposition is unique, up to multiplication by a unitary diagonal matrix, the matrix $Q$ in (3.2) must be the product of a unitary diagonal matrix with $\hat{H}$. Matrices of this form have eigenvalues uniformly distributed about the unit circle.

In contrast, Fig. 3a shows the eigenvalues of the unitary matrix $B$ obtained from a GMRES computation with a diagonal matrix $A$ of order $n = 19$ whose eigenvalues are uniformly distributed between 1 and 10, again using a random initial residual. Note the large gaps in the spectrum of this unitary matrix. The solid line in Fig. 3b shows the convergence of the GMRES algorithm, applied to either $A$ or $B$, with this same initial residual, while the dotted line in the figure shows the convergence of the GMRES algorithm applied to $B$ with a different random initial residual, uncorrelated with $B$. Again, the GMRES algorithm behaves similarly, when applied to either $A$ or $B$, with an arbitrary initial vector. Now the convergence is faster, which is characteristic of the GMRES algorithm applied to either a well-conditioned symmetric positive definite matrix or a unitary matrix with large gaps in its spectrum.

Also shown in Fig. 3b are sharp upper bounds for the residual at each step of the GMRES algorithm applied to $A$ (+) and to $B$ (x). Again, the sharp bounds for $A$ and $B$ are similar, and each is not much worse than the typical behavior of the algorithm for that matrix.

Although it seems surprising at first, it is perhaps not totally unexpected that any behavior that can be seen with the GMRES method can be seen with the method applied to a unitary matrix. It is known that the worst possible behavior – no progress at all until step $n$ – can occur with a unitary shift matrix [1,14], (this will be discussed further in section 4), and, of course, the best possible behavior – convergence in one step – occurs with the identity. We have shown that any convergence behavior between these two extremes can also occur with a unitary matrix.

**3.2. Equivalent Hermitian Positive Definite Matrices.** If zero is outside the field of values of $A$, then zero will also be outside the field of values of $H = W^*AW$. In this case, $H$ can be factored in the form $H = UL$, where $U$ and $L$ are nonsingular upper and lower triangular matrices, respectively. This decomposition can also be written in the form

$$H = (UL^{-*})(L^*L), \tag{3.4}$$

where the first factor $UL^{-*}$ is upper triangular and the second is Hermitian positive definite (and also tridiagonal). Thus, in this case, the GMRES method applied to $A$ behaves just as it does when applied to a certain Hermitian positive definite matrix!

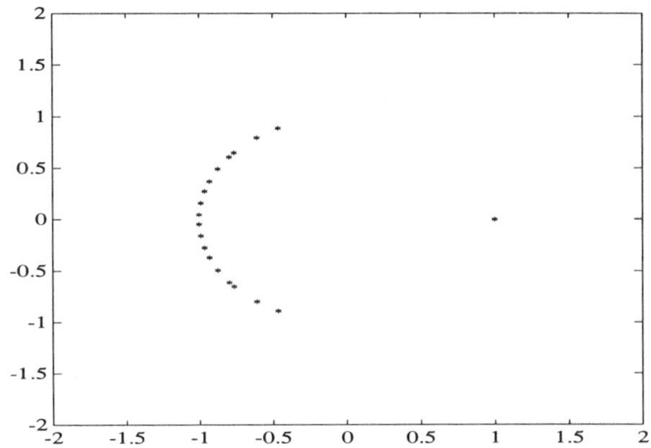


Fig. 3a. Eigenvalues of B. A=diag(1,1.5,2, ...,10)

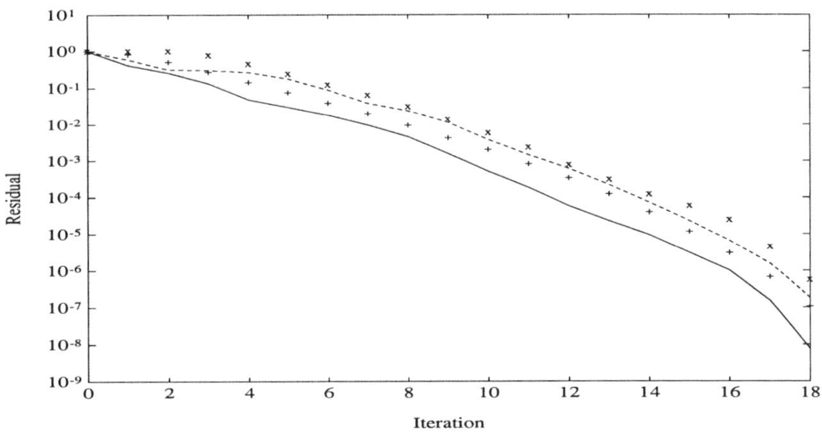


Fig. 3b. GMRMES(A or B,r0)(-) and GMRES(B,random)(–)

$$GMRES(A, r^0) \equiv GMRES(WL^*LW^*, r^0). \tag{3.5}$$

Since $B = WL^*LW^*$ is positive definite, the GMRES algorithm converges strictly monotonically. This gives a new proof of the known fact that if zero is outside the field of values of $A$, then the GMRES method converges strictly monotonically. (See [3] or [4] for a different proof.)

If, in addition to having its field of values separated from the origin, the matrix $A$ is close to Hermitian, then $H = W^*AW$ will be close to Hermitian, and one might expect that if the matrices $U$ and $L$ in (3.4) are scaled properly, then the Hermitian matrix $L^*L$ would be close to $H = UL$. In this case the eigenvalues of $L^*L$ would be close to either eigenvalues or singular values of $A$, indicating that either of these quantities would essentially determine the convergence rate of the GMRES algorithm. We will not go through a formal perturbation analysis here (for such an analysis, see [17]), but make only the following simple observation about the condition number $\kappa$ of the matrix $L^*L$:

$$\kappa(L^*L) \leq \kappa(L^*U^{-1}) \cdot \kappa(H) \tag{3.6}$$

Any conditions on $A$ that would ensure that, for any unitarily similar upper Hessenberg matrix $H$, the properly scaled factors of the $UL$ decomposition of $H$ satisfy $\kappa(L^*U^{-1}) < \kappa(H)$, would ensure that the condition number of $B = WL^*LW^*$ is less than that of $A^*A$. If the eigenvalues of $A^*A$ are fairly uniformly distributed throughout the interval, this would then imply that the GMRES algorithm applied to $A$ converges in fewer iterations than the algorithm applied to the normal equations $A^*A$ (though, of course, the comparison of total work would depend on the number of steps taken and the cost of applying $A$ and $A^*$). Unfortunately, we do not know of simple and interesting conditions on the matrix $A$ that guarantee this property.

One can derive rough bounds on the size of the $U$ and $L$ factors separately, based on the distance from the field of values of $A$ to the origin. Similar bounds on the size of $U^{-1}$ and $L^{-1}$ can be expressed in terms of the distance from the field of values of $A^{-1}$ to the origin. If both of these distances are fairly large, then we believe that the typical or worst-case behavior of the GMRES algorithm applied to $L^*L$ will be similar to that of the algorithm applied to $A$. This is in contrast to the case in which the field of values of $A$ contains the origin. For such problems, it may still be possible to factor the matrix $H$ in the form (3.4) but now the matrix $L^*L$ may be arbitrarily ill-conditioned. In such cases, the typical behavior of the GMRES algorithm applied to $L^*L$ may be much worse than that of the method applied to $A$, even though the behavior is identical for the particular initial vector $r^0$ used in generating $L^*L$.

As an example, we considered the matrix of Lenferink and Spijker [12]. This is a non-normal tridiagonal matrix of the form

$$\text{tridiag}[\ 1/(i+1),\ -3-2i,\ i+1\ ], \quad i = 1, ..., n.$$

It is strongly diagonally dominant and its field of values is well-separated from the origin [19]. Rather than using a random initial vector this time, for each step $k$, we determined an initial residual $r^0$ that gave rise to a residual of largest possible norm at step $k$. This was done using an optimization code, as described previously. For each of these initial vectors, we then computed the Hessenberg matrix $H$ and the factorization (3.4), where $U$ and $L$ were scaled to have the same diagonal elements. The goal was to determine how ill-conditioned $L^*L$ is for these "worst-case" initial vectors.

We used a matrix $A$ of order $n = 16$. The condition number of $A$ was 10.6. The condition number of $L^*L$ ranged from 15.6 to 27.3 – greater than that of $A$, but less than that of $A^*A$.

Fig. 4 shows the residual bound

$$(3.7) \qquad 2\left[\left(\frac{\sqrt{\kappa}-1}{\sqrt{\kappa}+1}\right)^k + \left(\frac{\sqrt{\kappa}+1}{\sqrt{\kappa}-1}\right)^k\right]^{-1}$$

based on the most ill-conditioned matrix $L^*L$ , with $\kappa = 27.3$ (solid line). From (3.5), this is also a bound on the convergence rate of the GMRES method applied to $A$. The dashed line in Fig. 4 shows the sharp error bound, which was computed numerically. Note the similarity between the two, at least in the early steps of the computation. This suggests that if a reasonable bound on the condition number of $L^*L$ could be established a priori, then (3.7) would give a realistic bound on the convergence rate of the GMRES algorithm applied to $A$.

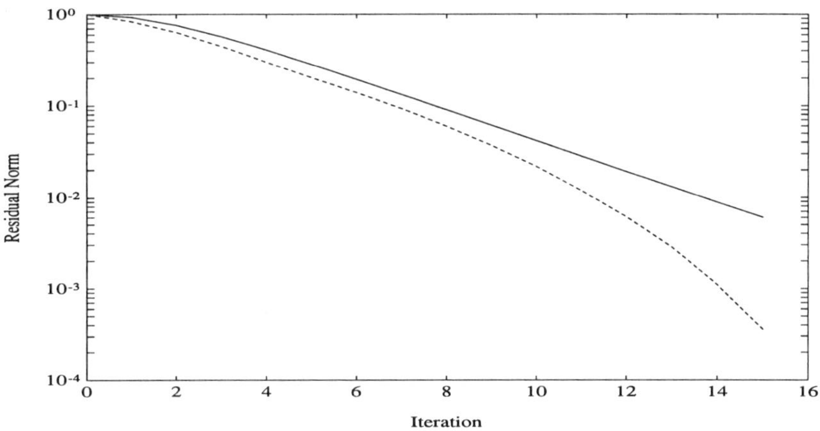


Fig. 4. Bound from HPD Matrix (-) and Sharp Bound (–)

**3.3. Equivalent Hermitian Matrices.** Using the characterization of Theorem 3, for most matrices $A$ and initial residuals $r^0$, it is possible to define a Hermitian (not necessarily positive definite) matrix $B$ such that $GMRES(A, r^0) \equiv GMRES(B, r^0)$. If the matrix $H^{-1}$ of Theorem 3 has no zero columns in its lower triangle, then let the Hermitian matrix $X$ be defined by

$$(3.8) \qquad X = \mathcal{L}(H^{-1})D + (\hat{\mathcal{L}}(H^{-1})D)^*,$$

where $D$ is any diagonal matrix such that $\mathcal{L}(H^{-1})D$ has real diagonal elements. (Recall that $\mathcal{L}(H^{-1})$ denotes the lower triangle of $H^{-1}$ including the diagonal, while $\hat{\mathcal{L}}(H^{-1})$ denotes the strict lower triangle of $H^{-1}$. We will also use the notation $\mathcal{U}(\cdot)$ and $\hat{\mathcal{U}}(\cdot)$ to denote the upper triangle and strict upper triangle of a matrix, respectively.) If $H^{-1}$ has a column of zeros in its lower triangle, say, column $j$, then that column can be replaced by any column of the form (2.13) provided the diagonal element $X_{jj}$ is real, and then the strict upper triangle of $X$ can be taken to be the Hermitian transpose of the strict lower triangle. If the diagonal matrix $D$ in (3.8), or, the elements $X_{nj}$ in (2.13) can be chosen so that the Hermitian matrix $X$ is nonsingular, then, according to Theorem 3, $H$ can be written in the form $H = RX^{-1}$ for some nonsingular upper triangular matrix $R$, and so we have

$$GMRES(A, r^0) \equiv GMRES(WX^{-1}W^*, r^0). \tag{3.9}$$

If the matrix $A^{-1}$ is close to Hermitian, then $H^{-1} = W^*A^{-1}W$ will be close to Hermitian, and one might expect that the Hermitian matrix $X$ could be chosen to be close to $H^{-1}$ and hence to have nearby eigenvalues. From this it would follow that the inverses of the eigenvalues of $B = WX^{-1}W^*$ would be close to the inverses of the eigenvalues of $A$. We will quantify this statement, limiting ourselves, for simplicity, to the case of real matrices $A$ and real initial residuals $r^0$.

Suppose $A^{-1}$ is real and close to symmetric, say,

$$A^{-1} = M + E$$

where $M$ is symmetric and $\|E\|_F \leq \delta$. [We can take $M$ to be the symmetric part of $A^{-1}$, $\frac{1}{2}(A^{-1} + A^{-T})$ , since this is the closest symmetric matrix to $A^{-1}$ in the Frobenius norm (see, e.g. [10]), and then $E$ will be the skew-symmetric part, $\frac{1}{2}(A^{-1} - A^{-T})$ ]. Then $H^{-1}$ is also close to symmetric:

$$H^{-1} = W^T M W + W^T E W = \tilde{M} + F$$

where $\tilde{M} = W^T M W$ is symmetric and $\|F\|_F = \|W^T E W\|_F \leq \delta$. If we define $X$ using (3.8) with $D$ equal to the identity, then the difference between $X$ and $H^{-1}$ is given by

$$X - H^{-1} = \mathcal{L}(\tilde{M}) + \mathcal{L}(F) + \hat{\mathcal{U}}(\tilde{M}) + (\hat{\mathcal{L}}(F))^T - \tilde{M} - F = (\hat{\mathcal{L}}(F))^T - \hat{\mathcal{U}}(F),$$

and this satisfies

$$\|X - H^{-1}\| \leq \|X - H^{-1}\|_F \leq \sqrt{2}\, \|F\|_F \leq \sqrt{2}\, \delta. \tag{3.10}$$

Since $X$ is symmetric, this implies that the eigenvalues of $X$ differ from those of $H^{-1}$ by no more than $\sqrt{2}\delta$, or, equivalently, that the eigenvalues of $B^{-1} = WXW^T$ differ from those of $A^{-1}$ by no more than $\sqrt{2}\delta$. If we make the additional assumption that the absolute value of the smallest eigenvalue of $H^{-1}$ is greater than $\sqrt{2}\delta$, then this will ensure that the matrix $X$ is nonsingular. Since $\delta$ is a bound on the size of the skew-symmetric part of $A$, a sufficient assumption in terms of $A$ and its eigenvalues $\lambda_1, ..., \lambda_n$ is

$$\min_{i=1,\ldots,n} \frac{1}{|\lambda_i|} > \sqrt{2}\, \frac{\|A^{-1} - A^{-T}\|_F}{2}, \tag{3.11}$$

With the assumption (3.11), then, the residual bound (1.7) can be replaced by

$$\frac{\|r^k\|}{\|r^0\|} \leq \min_{p_k \epsilon \mathcal{P}_k} \max_{z^{-1} \epsilon \bigcup_{i=1}^{n} \mathcal{B}(\lambda_i^{-1}, \sqrt{2}\delta) \bigcap \mathbf{R}} |p_k(z)|, \tag{3.12}$$

where $\mathcal{B}(\lambda_i^{-1}, \sqrt{2}\delta)$ denotes the ball of radius $\sqrt{2}\delta$ centered at $\lambda_i^{-1}$ and $\mathbf{R}$ denotes the real numbers.

If the eigenvalues of $A^{-1}$ are sufficiently well-separated that the disks $\mathcal{B}(\lambda_i^{-1}, \sqrt{2}\delta)$ do not overlap, then $X$ will have exactly one eigenvalue in each of these disks, and if the eigenvalues are ordered accordingly we can write

$$\frac{\|r^k\|}{\|r^0\|} \leq \max_{\{\mu_1,\ldots,\mu_n:\ |\frac{1}{\mu_i} - \frac{1}{\lambda_i}| \leq \sqrt{2}\delta,\ \mu_i \epsilon \mathbf{R},\ i=1,\ldots,n\}} \min_{p_k \epsilon \mathcal{P}_k} \max_{i=1,\ldots,n} |p_k(\mu_i)|. \tag{3.13}$$

The bounds (3.12) and (3.13) are frequently overestimates of the actual residual, because the bound $\sqrt{2}\delta$ on the distance from each eigenvalue of $X$ to that of $A^{-1}$ is a fairly large overestimate. Still, for matrices that are close to symmetric but have ill-conditioned eigenvectors, the bounds (3.12) and (3.13) may be much smaller than (1.7). For such problems they are often comparable to the pseudo-spectral bound (1.11), for the best value of $\epsilon$.

**4. Eigenvalue information alone cannot ensure fast convergence.** We have mentioned several cases in which the eigenvalues of $A$ essentially determine the convergence rate of the GMRES algorithm:

1. If the matrix $Z$ of eigenvectors is well-conditioned, then the bound (1.7) is nearly sharp.
2. If, for some fairly small value of $\epsilon$, the length of the boundary of the $\epsilon$-pseudo-spectrum is not much greater than $2\pi\epsilon$, then (1.11) gives a reasonable bound on the size of the residual.
3. If $A$ is close to Hermitian in the sense of (3.11), then the bound (3.12) or (3.13) is a reasonable estimate.

In general, however, eigenvalue information alone cannot ensure fast convergence of the GMRES algorithm. Any behavior that can be seen with the method can be seen with the method applied to a matrix having *any* nonzero eigenvalues. This is most easily seen from Theorem 2. Let $C$ be a companion matrix with the desired eigenvalues, say,

$$C = \begin{pmatrix} 0 & . & . & 0 & \alpha_0 \\ 1 & & & & \alpha_1 \\ & . & & & . \\ & & . & & . \\ & & & 1 & \alpha_{n-1} \end{pmatrix}. \tag{4.1}$$

(The eigenvalues of this matrix are the roots of the polynomial $z^n - \sum_{j=0}^{n-1} \alpha_j z^j$, and the constants $\alpha_j$, $j = 0, ..., n-1$ can be chosen so that $C$ has the desired eigenvalues.) Define the upper triangular matrix $\hat{R}$ in Theorem 2 by

$$\hat{R} = \begin{pmatrix} 1 & 0 & . & 0 & \alpha_1 + < v, w^1 > \alpha_0 \\ 0 & 1 & & . & \alpha_2 + < v, w^2 > \alpha_0 \\ . & & . & . & . \\ 0 & . & . & 1 & \alpha_{n-1} + < v, w^{n-1} > \alpha_0 \\ 0 & 0 & 0 & 0 & < v, w^n > \alpha_0 \end{pmatrix}.$$

Then it is easily seen that $\hat{H}\hat{R}$ is equal to $C$, and since the spectrum of $\hat{R}\hat{H}$ is the same as that of $\hat{H}\hat{R}$, the matrix $B = W\hat{R}\hat{H}W^*$ also has the desired spectrum. We have thus proved the following corollary to Theorem 2:

COROLLARY 4.1. *Given any nonzero values $\mu_1, ..., \mu_n$, there is a matrix $B$ with eigenvalues $\mu_1, ..., \mu_n$ such that (2.4) holds.*

As an example, let us consider matrices that give rise to the worst possible behavior – no progress at all until step $n$ – when the initial residual is the first unit vector $e_1$. The following unitary shift matrix was given in ([1]) and ([14]) to illustrate this phenomenon:

$$A = \begin{pmatrix} 0 & 1 & 0 & \dots & 0 \\ 0 & 0 & 1 & \dots & 0 \\ . & . & . & . & . \\ . & . & . & . & . \\ 0 & 0 & 0 & \dots & 1 \\ 1 & 0 & 0 & \dots & 0 \end{pmatrix}.$$

It is clear that the initial residual $e_1 = (1, 0, ..., 0)^T$ is already orthogonal to the space $[Ae_1, ..., A^{n-1}e_1]$, and so the residual is not changed until step $n$. The Hessenberg matrix $H$ and the unitary matrix $W$ generated by the GMRES algorithm with this initial vector are

$$H = \begin{pmatrix} 0 & 0 & \dots & 0 & 1 \\ 1 & 0 & \dots & 0 & 0 \\ 0 & 1 & \dots & 0 & 0 \\ . & . & . & . & . \\ . & . & . & . & . \\ 0 & 0 & \dots & 1 & 0 \end{pmatrix}, \quad W = \begin{pmatrix} 0 & 0 & \dots & 0 & 1 \\ 0 & 0 & \dots & 1 & 0 \\ . & . & . & . & . \\ . & . & . & . & . \\ 0 & 1 & \dots & 0 & 0 \\ 1 & 0 & \dots & 0 & 0 \end{pmatrix}.$$

According to Theorem 1, the class of matrices that generate the same Krylov residual spaces as $A$ (with initial vector $e_1$) consists of all matrices of the form $WRHW^*$, where $R$ is any nonsingular upper triangular matrix; i.e., all nonsingular matrices of the form

$$\begin{pmatrix} 0 & * & 0 & \dots & 0 \\ 0 & * & * & \dots & 0 \\ . & . & . & . & . \\ . & . & . & . & . \\ 0 & * & * & \dots & * \\ * & * & * & \dots & * \end{pmatrix},$$

where the $*$'s can be any values.

This class includes, for example, all companion matrices of the form $C^*$, where $C$ is given in (4.1). As noted previously, the constants $\alpha_j$, $j = 0, \dots, n-1$, can be chosen to make this matrix have any desired eigenvalues, say, all 1's. It follows that the worst-case behavior of the GMRES algorithm (i.e., no progress until step $n$) can occur for a matrix having what one might have expected to be the best possible eigenvalues!

**5. Conclusions, open questions, and possible applications.** We have pointed out several cases in which the eigenvalues of the matrix $A$ essentially determine the behavior of the GMRES algorithm. We have shown, in general, however, that this is not the case. The same behavior can be seen for a matrix having any given eigenvalues.

We have introduced a technique for analyzing the behavior of the GMRES algorithm applied to non-normal matrices by defining an equivalence class containing normal matrices – the class of matrices that, with a given initial residual, produce the same GMRES residuals at every step. In this regard, we have raised more questions than we have answered. We have shown that there is always an equivalent unitary matrix. We have demonstrated numerically that certain properties of the eigenvalues of this unitary matrix are related to special properties of $A$, but we have not proved such a relationship. We have shown that if zero is outside the field of values of $A$ then there is an equivalent Hermitian positive definite matrix, but we have not been able to bound the condition number of this Hermitian positive definite matrix in terms of simple properties of $A$.

Our analysis still has not explained the behavior of the first example given in Section 1 – a matrix $A$ of the form $Z\Lambda Z^{-1}$, where $Z$ and $\Lambda$ are defined by (1.12). There we claimed that the eigenvalues of this matrix essentially determine the convergence of the GMRES iteration. Yet, for certain initial residuals, there can be no normal matrix in $A$'s equivalence class whose eigenvalues are close to those of $A$. This is because the field of values of $A$ contains the origin. It follows that there is an initial vector for which the GMRES iteration makes no progress at step one. Any normal matrix with this behavior must have eigenvalues on both sides of the origin, and so its spectrum cannot be close to the positive spectrum of $A$. Using several different random initial residuals and also the initial residuals that give rise to the largest possible residuals at steps one and two, we computed the equivalent unitary matrix $WQW^*$ defined by (3.2) and the equivalent Hermitian matrix $WX^{-1}W^*$ defined by (3.8) (with $D = I$). In all cases, there were large gaps in the spectrum of the unitary matrix, similar to those shown in Fig. 3a. The equivalent Hermitian matrix had one large negative eigenvalue (which probably was not computed accurately due to rounding errors) and the remaining eigenvalues were close to those of $A$ – all but one were in the interval $[1, 5]$ and the remaining one was less than 20. This indicates that for this problem as well, there are normal matrices in $A$'s equivalence class for which the typical behavior of the GMRES algorithm is similar to that for $A$. It remains an open question how to show a priori that this is the case. Another possible approach is to consider near-normal matrices in the equivalence class – matrices with well-conditioned (but not necessarily unitary) eigenvector matrices. There may be a near-normal matrix in $A$'s equivalence class whose eigenvalues are all close to those of $A$.

The equivalence class we have defined consists of matrices for which the residuals at each step of the GMRES algorithm are identical. In practice, one seldom runs a GMRES computation for a full $n$ steps. Instead, one runs for some fixed number of steps, say, $m << n$, and then restarts. The class of matrices that generate the same residuals at steps 1 through $m$ is broader than the class we have considered. Investigation of this larger class may lead to results about the convergence of the restarted GMRES iteration.

This technique of analysis – equating the behavior of the algorithm applied to a given problem to its behavior when applied to some other problem that is better understood – may prove useful for analyzing other iterative methods as well. The biconjugate gradient [5] and QMR [7,6] iterations are examples, and work along these lines has begun.

Finally, we would like to note that the results of Section 4 are somewhat discouraging, as far as Krylov space iterative methods are concerned. It is disappointing that one can determine, just from the sparsity pattern of a matrix, that, for some initial residual, every Krylov space method will converge poorly (or diverge)! Similar behavior can be expected with a random initial residual, as our numerical experiments indicate. While it is sometimes possible to choose an initial guess for which the algorithm will display better convergence, we believe that *typically* the initial guess plays only a small role in determining the convergence rate of the algorithm, and so this behavior is to be expected for problems with such coefficient matrices. (And, of course, this same behavior would be observed for any matrix unitarily similar to one with a bad sparsity pattern.) This suggests that it is important to consider methods that choose approximate solutions from outside the Krylov space. An example is the GMRESR method of van der Vorst and Vuik [20]. Another approach is to look for preconditioners that make the preconditioned matrix close to normal, in some appropriate sense. Further research along these lines would seem an important next step.

**Acknowledgment.** We would like to thank Nick Trefethen for helpful discussions and advice about the subject matter of this paper.

## REFERENCES

[1] P. Brown, "A Theoretical Comparison of the Arnoldi and GMRES Algorithms," SIAM Jour. Sci. Stat. Comput. 20 (1991), pp. 58-78.

[2] N. Dunford and J. T. Schwartz, *Linear Operators. Part I: General Theory*, Interscience Publishers, Inc., New York, 1958.

[3] S. C. Eisenstat, H. C. Elman, and M. H. Schultz, "Variational Iterative Methods for Nonsymmetric Systems of Linear Equations," SIAM J. Num. Anal. 20 (1983), pp. 245-357.

[4] H. C. Elman, "Iterative Methods for Large, Sparse, Nonsymmetric Systems of Linear Equations," Ph.D. thesis and Res. Rep. No. 229, Dept. of Comp. Sci., Yale U., 1982.

[5] R. Fletcher, "Conjugate Gradient Methods for Indefinite Systems," in G. A. Watson, ed., Proc. of the Dundee Biennial Conference on Numerical Analysis, Springer-Verlag, New York, 1975.

[6] R. Freund, M. Gutknecht, and N. Nachtigal, "An Implementation of the Look-Ahead Lanczos Algorithm for Non-Hermitian Matrices, " RIACS Tech. Report 91.09, April, 1991.

[7] R. Freund and N. Nachtigal, "QMR: a Quasi-Minimal Residual Method for Non-Hermitian Linear Systems," RIACS Tech. Report 91.09, April, 1991.

[8] A. Greenbaum and L. Gurvits, "Max-Min Properties of Matrix Factor Norms," to appear.

[9] A. Greenbaum and L. N. Trefethen, "GMRES-CR and Arnoldi-Lanczos as Matrix Approximation Problems," to appear.

[10] T. Hayden and J. Wells, "Approximation by Matrices Positive Semidefinite on a Subspace," Lin. Alg. Appl. 109 (1988), pp. 115-130.

[11] W. Joubert, "A Robust GMRES-Based Adaptive Polynomial Preconditioning Algorithm for Nonsymmetric Linear Systems," to appear.

[12] H. W. J. Lenferink and M. N. Spijker, "A Generalization of the Numerical Range of a Matrix," Lin. Alg. Appl. 140 (1990), pp. 251-266.

[13] C. Moler, J. Little, and S. Bangert, *PRO-MATLAB for Sun Workstations*, The MathWorks, Inc., Sherborn, MA, August, 1987.

[14] N. M. Nachtigal, S. Reddy, and L. N. Trefethen, "How Fast are Nonsymmetric Matrix Iterations?" Proceedings of the Copper Mountain Conference on Iterative Methods, Copper Mountain, CO, 1990.

[15] S. C. Reddy, "Pseudospectra of Operators and Discretization Matrices and an Application to Stability of the Method of Lines," Ph.D. thesis, MIT, 1991.

[16] Y. Saad and M. H. Schultz, "GMRES: A Generalized Minimum Residual Algorithm for Solving Nonsymmetric Linear Systems," SIAM J. Sci. Stat. Comput. 7 (1986), pp. 856-869.

[17] G. W. Stewart, "On the Perturbation of LU, Cholesky, and QR Factorizations," in preparation.

[18] L. N. Trefethen, "Approximation Theory and Numerical Linear Algebra," in *Algorithms for Approximation II*, J. C. Mason and M. G. Cox, eds., Chapman and Hall, London, 1990.

[19] L. N. Trefethen, "Pseudospectra of Matrices," in D. F. Griffiths and G. A. Watson, Numerical Analysis 1991, Longman, 1992.

[20] H. van der Vorst and C. Vuik, "GMRESR: A Family of Nested GMRES Methods," Report 91-80, Delft University of Technology, 1991.

[21] P. Vinsome, "Orthomin, An Iterative Method for Solving Sparse Sets of Simultaneous Linear Equations," SPE 5729, Fourth Symposium on Numerical Simulation of Reservoir Performance, SPE, Feb., 1976.

[22] D. M. Young and K. C. Jea, "Generalized Conjugate Gradient Acceleration of Nonsymmetrizable Iterative Methods," Lin. Alg. Appl. 34 (1980), pp. 159-194.

# INCOMPLETE BLOCK FACTORIZATIONS AS PRECONDITIONERS FOR SPARSE SPD MATRICES

L.YU. KOLOTILINA* AND A.YU. YEREMIN †

**Abstract.** The paper deals with incomplete block factorization (IBF) preconditionings for block tridiagonal SPD matrices. Two schemes of IBF's (the standard one and the one in inverse free form) are considered and shown to give rise to SPD preconditioners under certain general assumptions on sparse approximations to pivot blocks. An implementation of CG iterations preconditioned with an IBF in inverse free form on a two-dimensional grid of processors is suggested. Performance of IBF-CG iterations is demonstrated by numerical results for 3D linear elasticity problems.

**1. Introduction.** In the last decade much attention has been paid to developing Incomplete Block Factorizations (IBF's) and to justifying their use as preconditioners for the Conjugate Gradient (CG) type methods, see, e.g., [1, 3, 4, 6, 9]. However, as far as we are aware, the whole existence and convergence theory for IBF's has been based on diagonal dominance arguments and dealt with M-, H-, or block H-matrices.

The present analysis concerns general Symmetric Positive Definite (SPD) matrices partitioned into a block tridiagonal form and is applicable to off-diagonally dominant matrices. We consider the two commonly used types of IBF's and show that under certain general assumptions on sparse approximations to pivot blocks and their inverses both types of IBF's exist while eigenvalues of the corresponding preconditioned matrices lie in the interval (0,1]. We also provide examples of specific sparse approximations that satisfy the required conditions which makes our theory applicable in practice.

The paper is organized as follows. In the next section we present the existence theory for the so-called standard and inverse free forms of IBF's. Section 3 describes sparse approximation methods applied to pivot blocks and their inverses. In Section 4 we suggest a scheme for multiplying a block SSOR type preconditioned matrix by a vector generalizing that in [8] which is computationally effective if the block off-diagonal part of the matrix under preconditioning is not too sparse. We also present an implementation of CG iterations preconditioned with an inverse free IBF for 3D finite element problems on a two-dimensional grid of processors. Communication overhead of this implementation scheme depends linearly on the length of block recursion and is a decreasing function of the number of processors used (except for inner product computations). This implies that the amount both

* St. Petersburg Branch of V.A., Steklov Mathematical Institute of the Russian Academy of Sciences, Fontanka 27, St. Petersburg 191 011, Russia

† Institute of Numerical Mathematics of the Russian Academy of Sciences, Leninskij prosp. 32A, Moscow 117 334, Russia. E-mail: liko@lomi.spb.su and badger@adonis.ias.msk.su

of overhead in exploiting parallelism and of serial work grows much slower than the amount of parallel work when the problem size increases and thus for larger problems more processors can be efficiently utilized. The next section contains the results of numerical experiments demonstrating performance of IBF-CG methods for finite element discretizations of a 3D linear elasticity problem. Section 6 presents some conclusions.

**2. Incomplete block factorization preconditionings for block tridiagonal SPD matrices.** In this section for block tridiagonal SPD matrices we prove existence and positive definiteness of incomplete block factorizations in the so-called standard and inverse free forms under general assumptions on the approximations to pivot blocks and their inverses and show that under certain conditions eigenvalues of the IBF preconditioned matrices do not exceed 1.

Let $A = (A_{ij})$ be an $n \times n$ block tridiagonal (i.e. $A_{ij} = 0$ for $|i-j| > 1$) SPD matrix presented also in the form

$$A = D + L + L^T,$$

where $D = Diag(A) = Diag(D_1 \ldots, D_n)$ is the block diagonal part of $A$, while $L$ is its strictly lower block triangular part.

Since $A$ is a SPD block tridiagonal matrix its exact block factorization, as is well known, exists and can be written in the form

$$A = (G^{-1} + L)G(G^{-1} + L^T),$$

where $G = Diag(G_1, \ldots, G_n)$ satisfies the relations

$$\begin{aligned} &G_1 = D_1^{-1}, \\ &G_1 = (D_i - A_{ii-1}G_{i-1}A_{i-1i})^{-1}, i \geq 1. \end{aligned} \tag{2.1}$$

The incomplete block factorization of $A$ in inverse free form introduced in [1] and studied, e.g., in [3,6,9] is then constructed in the form similar to (2.1)

$$B = (F^{-1} + L)F(F^{-1} + L^T) = (I + L\ F)F^{-1}\ (I + F\ L^T), \tag{2.2}$$

where $F = Diag(F_1, \ldots, F_n)$ is a block diagonal matrix whose blocks are determined using the recurrence relations

$$\begin{aligned} &F_i = \Omega_i(X_i^{-1}), \quad i \geq 1, \\ &X_1 = D_1, \\ &X_i = D_i - A_{ii-1}F_{i-1}A_{i-1i}, \quad i \geq 2, \end{aligned} \tag{2.3}$$

where by $\Omega_i(X_i^{-1})$ we denote a sparse approximation of the inverse to $X_i$. Clearly, for the incomplete factorization (2.2)–(2.3) to be SPD all $F_i$ must be SPD and so we must assume that the approximation rules $\Omega_i$ are such that symmetry and positive definiteness of a matrix $X$ are preserved in the matrices $\Omega_i(X^{-1})$ approximating its inverse.

We are now ready to formulate our first result.

THEOREM 2.1. *Let $A$ be a SPD block tridiagonal matrix and approximation rules $\Omega_i$, $1 \leq i \leq n-1$, be such that for any SPD matrix $X$ (of the corresponding order) the matrices $\Omega_i(X^{-1})$ are symmetric and the inequalities (in the sense of quadratic forms)*

$$0 < \Omega_i(X^{-1}) \leq X^{-1} \tag{2.4}$$

*are valid. Then the incomplete block factorization (2.2) – (2.3) exists, $B$ is a SPD matrix, and $Spec(B^{-1}A) \subset (0,1]$, where by $Spec(B^{-1}A)$ we denote the spectrum of the matrix $B^{-1}A$.*

**Proof.** We proceed by induction. Clearly, for $i=1$ we have by (2.1) and (2.3) that $G_1 = D_1^{-1}$ and $F_1 = \Omega_1(D_1^{-1})$. Therefore, by (2.4)

$$0 < F_1 \leq G_1.$$

Assuming now the inequalities

$$0 < F_i \leq G_i \tag{2.5}$$

to be already proved for $i = 1, \ldots, k-1$, let us prove them for $i = k$. By (2.3), (2.5), and (2.1) we have

$$\begin{aligned} X_k &= D_k - A_{kk-1}F_{k-1}A_{k-1k} \\ &\geq D_k - A_{kk-1}G_{k-1}A_{k-1k} = G_k^{-1} \end{aligned}$$

implying that $X_k$ is a SPD matrix and $X_k^{-1} \leq G_k$. Hence, by (2.4)

$$0 < F_k = \Omega_k(X_k^{-1}) \leq X_k^{-1} \leq G_k \tag{2.6}$$

and thus the inequalities (2.5) showing that $B$ exists and is SPD are proved for all i.

It remains only to show that eigenvalues of $B^{-1}A$ do not exceed 1. Indeed, using (2.6) and (2.3) we have

$$\begin{aligned} B &= F^{-1} + L + L^T + L\ F\ L^T \geq X + L + L^T + L\ F\ L^T \\ &= (D - L\ F\ L^T) + L + L^T + L\ F\ L^T = D + L + L^T = A \end{aligned}$$

and thus $B \geq A$ which completes the proof. □

Consider next the standard form of the incomplete block factorization of a block tridiagonal SPD matrix analysed in many papers, see, e.g, [1,3,4,6,9]. It is based on the exact block factorization written as

$$A = (\Delta + L)\Delta^{-1}(\Delta + L^T)$$

with the block diagonal matrix $\Delta$ satisfying the equality

$$\Delta = D - L\,\Delta^{-1}L^T \tag{2.7}$$

and is constructed in the form

$$C = (\nabla + L)\nabla^{-1}(\nabla + L^T), \tag{2.8}$$

where $\nabla = Diag(\nabla_1, \ldots, \nabla_n)$ is a block diagonal matrix satisfying the recurrence relations

$$\begin{aligned} \nabla_1 &= \Omega'_1(D_1), \\ \nabla_i &= \Omega'_i(D_i - A_{ii-1}\Omega_{i-1}(\nabla_{i-1}^{-1})A_{i-1i}), \quad i = 2, \ldots, n, \end{aligned} \tag{2.9}$$

where $\Omega'_i(X)$ is a SPD approximation of a SPD matrix $X$.

An additional approximation stage consisting in replacing

$$D_i - A_{ii-1}\Omega_{i-1}(\nabla_{i-1}^{-1})A_{i-1i}$$

by $\Omega'_i(D_i - A_{ii-1}\Omega_{i-1}(\nabla_{i-1}^{-1})A_{i-1i})$ in this scheme of incomplete block factorization might be needed since linear systems with $\nabla$ are to be solved at each preconditioned iteration. This is usually done using Cholesky decomposition of $\nabla$ and thus the sparsity pattern of $\nabla$ must be controlled in order to control the sparsity pattern of its Cholesky factor determining complexity of the factorization and solution stages.

Conditions ensuring existence and positive definiteness of $\nabla$ and hence of $C$ are formulated in the next

THEOREM 2.2. *Let $A$ be a SPD block tridiagonal matrix and the approximation rules $\Omega_i$ and $\Omega'_i$, $1 \le i \le n$, be such that for any SPD matrix $X$ (of the corresponding order) the matrices $\Omega_i(X^{-1})$ and $\Omega'_i(X)$ are symmetric and satisfy for $\xi_i \ge \gamma_i$ the inequalities (in the sense of quadratic forms)*

$$\Omega_i(X^{-1}) \le \gamma_i X^{-1}, \quad \gamma_i \ge 0, \tag{2.10}$$

$$\Omega'_i(X) \ \ge \xi_i X, \quad \xi_i > 0, \tag{2.11}$$

*Then the incomplete block factorization (2.8) – (2.9) exists, $C$ is a SPD matrix, and in the case $\gamma_i \le 1, \xi_i \ge 1$, $i = 1, \ldots, n$, $Spec(C^{-1}A) \subset (0, 1]$.*

**Proof.** We prove the theorem by induction. Obviously for $i = 1$ the matrix $\nabla_1 = \Omega'_1(D_1) = \Omega'_1(\Delta_1)$ is symmetric and positive definite and by (2.11), $\nabla_1 \ge \xi_1\Delta_1$.

Let the inequalities

$$\nabla_i \ge \xi_i\Delta_i \tag{2.12}$$

be valid for $i \leq k-1$. Then for $i = k$ we also have by (2.9) and (2.11) that

$$\begin{aligned} \nabla_k &= \Omega'_k(D_k - A_{kk-1}\Omega_{k-1}(\nabla_{k-1}^{-1})A_{k-1k}) \\ &\geq \xi_k(D_k - A_{kk-1}\Omega_{k-1}(\nabla_{k-1}^{-1})A_{k-1k}) \\ &\geq \xi_k(D_k - A_{kk-1}\Delta_{k-1}^{-1}A_{k-1k}) = \xi_k\ \Delta_k \end{aligned}$$

since (2.10) and the assumption $\xi_i \geq \gamma_i,\ i = 1, \dots, n$, imply that

$$\Omega_{k-1}(\nabla_{k-1}^{-1}) \leq \gamma_{k-1}\nabla_{k-1}^{-1} \leq \gamma_{k-1}\xi_{k-1}^{-1}\Delta_{k-1}^{-1} \leq \Delta_{k-1}^{-1}.$$

Thus $\nabla$ is a well-defined SPD matrix and hence so is $C$.

Now since in the case $\xi_i \geq 1$ by (2.11)

$$\nabla_i \geq D_i - A_{ii-1}\Omega_{i-1}(\nabla_{i-1}^{-1})A_{i-1i},$$

using (2.10) and the inequalities $\gamma_i \leq 1$ we derive

$$\begin{aligned} C &= \nabla + L + L^T + L\nabla^{-1}L^T \\ &\geq D - L\Omega(\nabla^{-1})L^T + L + L^T + L\nabla^{-1}L^T \\ &= A + L(\nabla^{-1} - \Omega(\nabla^{-1}))L^T \geq A. \end{aligned}$$

Therefore, $Spec(C^{-1}A) \subset (0,1]$ and the proof is completed. □

To make the two schemes of IBF's for block tridiagonal SPD matrices practical in the next section we present examples of approximation rules $\Omega$ and $\Omega'$ satisfying conditions of the above theorems.

**3. Sparse approximations to SPD matrices and their inverses.** Consider first a method for constructing sparse approximations of an $m \times m$ SPD matrix $X$ based on the diagonally compensated reduction of off-diagonal entries considered in [2]. For a selected set $S$,

$$\{(i,i), 1 \leq i \leq n\} \subseteq S \subseteq \{(i,j) : 1 \leq i,j \leq n\}$$

let a matrix $Y = (y_{ij})$ be defined as follows:

$$\begin{aligned} y_{ij} &= 0 \quad , \quad (i,j) \notin S, \\ y_{ij} &= x_{ij} \quad , \quad (i,j) \in S. \end{aligned}$$

Then the remainder $X - Y$ can be split as

$$X - Y = P_1 - P_2,$$

where both $P_1$ and $P_2$ are nonnegative, $P_1$ consists of all positive entries of $X - Y$ while $-P_2$ consists of all negative entries of $X - Y$. Select then entrywise positive weight vectors $v_1$ and $v_2$ and construct nonnegative diagonal matrices $D_1$ and $D_2$ using the equalities

$$D_i v_i = P_i v_i,\ i = 1,2,$$

implying that the matrices $D_i \pm P_i$ are positive semidefinite. Finally, set

$$\Omega'(X) = Y + D_1 + D_2.$$

Then

$$\Omega'(X) = X + (D_1 - P_1) + (D_2 + P_2) \geq X \tag{3.1}$$

and thus condition (2.11) of Theorem 2.2 is satisfied with $\xi = 1$.

A number of examples for which the so constructed approximation $\Omega'(X)$ is quite good (in the sense that the condition number of $[\Omega'(X)]^{-1}X$ is small) can be found in [2]. On the other hand, it is clear that for some choices of $S$ the diagonal correction matrix $D_1 + D_2$ might happen to be too large and as a result the approximation quality of $\Omega'(X)$ might be poor.

Note that in case when $Y$ consists of negative off-diagonal entries of $X$ the approximation $\Omega'(X)$ is a Stieltjes matrix and thus using such $\Omega'$'s when constructing a standard IBF we need to approximate inverses only to Stieltjes matrices.

Consider next the problem of constructing sparse approximations $\Omega(X^{-1})$ of the inverse to a SPD matrix $X$. The only method known to the authors which provides a SPD approximation to $X^{-1}$ for any SPD matrix $X$ without additive diagonal corrections consists in using the so-called factorized sparse approximate inverses introduced and studied in [7] in the general unsymmetric case. Let us recall in brief this construction.

Select first a lower triangular sparsity pattern $S_L \subseteq \{(i,j) : 1 \leq i \neq j \leq n\}$ such that $S_L \supseteq \{(i,j) : i < j\}$ and construct a lower triangular matrix $L = (\ell_{ij})$ of the sparsity pattern $S_L$ using the relations

$$\begin{array}{lcl} \ell_{ij} = 0 & , & (i,j) \in S_L, \\ (LX)_{ij} = \delta_{ij} & , & (i,j) \notin S_L, \end{array} \tag{3.2}$$

where $\delta$ is the Kronecker symbol. As shown in [7], the so constructed lower triangular matrix $L$ coincides up to a left diagonal scaling with the minimizer of

$$||I - ML_A||_F^2 = tr[(I - ML_A)(I - L_A^T M^T)]$$

over all matrices $M$ of the sparsity pattern $S_L$.

The factorized sparse approximate inverse to $X$ is then defined as

$$\Omega(X^{-1}) = L^T D^{-1} L, \tag{3.3}$$

where the diagonal matrix $D$ satisfies

$$D = Diag(LXL^T) = Diag(L)$$

and can be readily shown to be positive definite. Therefore, $\Omega(X^{-1})$ is symmetric and positive definite for any triangular sparsity pattern $S_L$.

Note that as explained in [7] equations (3.2) for computing $L$ decouple into independent systems of equations each determining nonzero entries in a row of $L$. Thus, the construction of the factorized sparse approximate inverse is comletely parallelizable.

To meet the conditions of Theorems 2.1 and 2.2 we next show how to modify the above construction of $\Omega(X^{-1})$ so that

$$\Omega(X^{-1}) \leq X^{-1} \tag{3.4}$$

in the sense of quadratic forms. Let

$$LXL^T = D - R.$$

Select an entrywise positive vector $v$ and construct the diagonal matrix $\widetilde{D}$ using the relation

$$\widetilde{D}v = |R|v,$$

where $|R|$ is the matrix of absolute values of the entries of $R$. Then setting

$$\hat{D} = D + \widetilde{D}$$

yields (since $\widetilde{D} + R$ is positive semidefinite)

$$\hat{D} = LXL^T + (\widetilde{D} + R) \geq LXL^T$$

implying that

$$\Omega(X^{-1}) = L^T \hat{D}^{-1} L \leq X^{-1}$$

as required.

Note that in the special case when $X$ is a Stieltjes or a generalized diagonally dominant matrix requirement (3.4) can be satisfied without additional computations simply by setting

$$\Omega(X^{-1}) = 1/2\ L^T D^{-1} L,$$

where $D = Diag(L)$, since in this case the splitting $LXL^T = D - R$ is convergent (see [7]), i.e., $\rho(I - D^{-1/2}LXL^T D^{-1/2}) < 1$ implying that $\Omega(X^{-1}) = 1/2 L^T D^{-1} L < X^{-1}$.

*Remark* 3.1. The above described scaling procedure ensuring that $\Omega(X^{-1}) \leq X^{-1}$ as required by the existence theory in practice turns out to be unnecessary and even undesirable since it frequently deteriorates the IBF preconditioning quality. If, however, positive definiteness gets lost in the course of constructing an IBF due to breaking the condition $\Omega_i(X_i^{-1}) \leq X_i^{-1}$ the situation can be usually helped by recomputing a current $\Omega_i(X_i^{-1})$ for a smaller set $S_L$.

**4. Parallel implementation of IBF-CG iterations.** In this section we suggest a scheme of parallel implementation of CG iterations preconditioned with an IBF in inverse free form on a two-dimensional grid of processors with nearest neighbor connections.

Let a linear system

$$Ax = b \tag{4.1}$$

with an $n \times n$ block tridiagonal SPD matrix

$$\begin{array}{l} A = (A_{ij}) = D + L + L^T, \\ A_{ij} = 0 \quad if \quad |i-j| > 1, \end{array} \qquad 1 \le i, j \le n, \tag{4.2}$$

be given. Consider an IBF preconditioned system

$$My = (I + LF)^{-1}b, \tag{4.3}$$

where

$$M = (I + LF)^{-1}A(F^{-1} + L^T)^{-1} \tag{4.4}$$

and

$$y = (F^{-1} + L^T)x.$$

To perform CG iterations for the preconditioned system (4.3) we need essentially (exept for vector operations and inner product computations) to multiply the matrix $M$ by a vector. This can be done according to the following scheme generalizing that for implementing block SSOR preconditioned CG iterations suggested in [8].

Taking into account that

$$A = (F^{-1} + L) + (F^{-1} + L^T) + (D - 2F^{-1})$$

we obtain

$$A(F^{-1} + L^T)^{-1} = (F^{-1} + L)(F^{-1} + L^T)^{-1} + I + (D - 2F^{-1})(F^{-1} + L^T)^{-1}$$

implying that

$$M = (I+L^T F)^{-1} + (I+LF)^{-1}[I+D(F^{-1}+L^T)^{-1} - 2(I+L^T F)^{-1}]. \tag{4.5}$$

This presentation of the matrix $M$ ensures that the product $Mf$, where $f = (f_1^T, ..., f_n^T)^T$ is a vector partitioned into block components according to the block partitioning of $A$, can be computed in

$$NZ(A) + 2comp(F) + 6N - N_1 - N_n \tag{4.6}$$

arithmetic operations, where $comp(F)$ is the number of arithmetic operations required to multiply $F$ by a vector, $N$ is the pointwise dimension of $A$, $N_1$ and $N_n$ are the orders of $A_{11}$ and $A_{nn}$, respectively. Indeed, let

$$u = (F^{-1} + L^T)^{-1} f \ , \quad v = (I + L^T F)^{-1} f \ , \quad w = Du. \tag{4.7}$$

Then since $u = Fv$, block components of the vectors $u, v$, and $w$ can be computed simultaneously using the formulae

$$\begin{array}{lll} v_n = f_n, & u_n = F_n v_n, & \\ for & i = n-1, ..., 1 & \\ & v_i = f_i - \sum_{j>i} A_{ij} u_j, & w_{i+1} = A_{i+1i+1} u_{i+1}, \\ & u_i = F_i v_i, & \\ w_1 = A_{11} u_1, & & \end{array} \tag{4.8}$$

which requires one multiplication of a vector by $L^T$, one by $D$, and one by $F$. Then, using (4.5) and (4.7) we have

$$Mf = v + (I + LF)^{-1}[f + w - 2v] \tag{4.9}$$

and thus up to vector operations the product $Mf$ can be computed using two matrix-vector multiplications with $F$, one with $D$, one with $L$, and one with $L^T$ which proves (4.6).

Note that the straightforward computation of the vector $Mf$ by solving first a block triangular system with the matrix $F^{-1} + L^T$, multiplying the result by $A$, and solving again a block triangular system with the matrix $I + LF$ requires

$$NZ(A) + 2comp(F) + 2NZ(L) + 2N - N_1 - N_n$$

arithmetic operations. Thus, multiplying the block SSOR type preconditioned matrix $M$ by a vector using (4.7) – (4.9) results in saving $2NZ(L) - 4N$ arithmetic operations per each CG iteration whenver the block off-diagonal part $L + L^T$ of $A$ is not too sparse.

To perform one CG iteration for (4.3) we obviously need to multiply blocks of A and F by vectors. Let us thus describe how individual blocks of $A$ and $F$ can be multiplied by vectors on a $k \times k$ grid of processors with nearest neighbor connections and duplex communication channels. To this end we assume that $A$ results from a finite element discretization of a 3D problem on a (topologically) cubic $s \times s \times s$ mesh with naturally ordered degrees of freedom, say, by horizontal planes from top to bottom. Let, furthermore, a block tridiagonal block partitioning of A be determined by assembling in the $i$'th diagonal block $A_{ii}$ rows and columns corresponding to the degrees of freedom associated with the $i$'th horizontal $s \times s$ plane mesh.

Then specifying a uniform partitioning of the corresponding $s \times s$ mesh into a $k \times k$ superelement mesh as shown in Fig. 1 induces a block $k \times k$

nine diagonal structure determined by the 2D nine point stencil on each of the blocks $A_{ij}$ and $F_i$ (under some restrictions on the sparsity pattern of $F$ dependent on the superelement size).

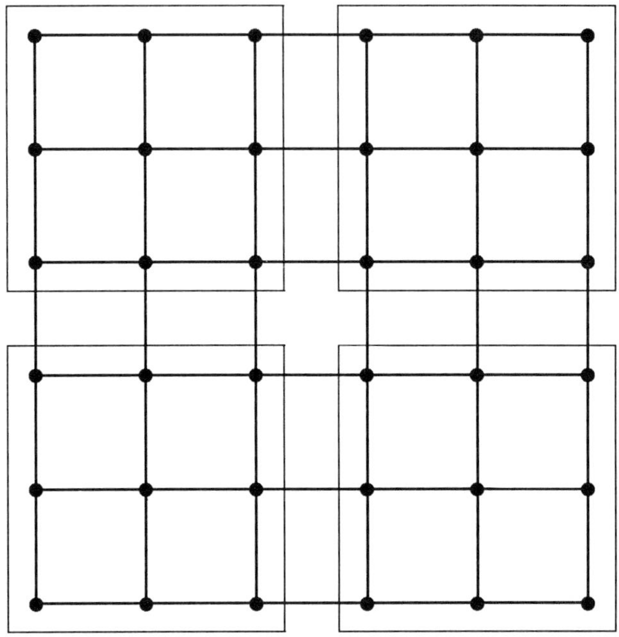

FIG. 4.1. *A* $2 \times 2$ *mesh of* $2 \times 2$ *non-overlapping superelements*

Associate each of the $k^2$ superelements with the processor having the same coordinates in the 2D grid of processors and allocate the $\ell$'th block rows, $1 \leq \ell \leq k^2$, of all the matrices $A_{ij}$ and $F_i$ corresponding to a superelement and the corresponding block components of the required vectors to the processor associated with this superelement. Then the multiplication of each of the blocks $A_{ij}$ and $F_i$ by a vector requires 2 data communications for transmitting block components of the vectors as shown in Fig. 2.

Note that with this data allocation the multiplication of the preconditioned matrix $M$ by a vector using either (4.7) – (4.9) or the straightforward approach requires $8n - 2$ data communications. Thus for the considered parallel implementation

- communication overhead in computing $M$ times a vector is proportional to the length of block recursion $n$ and thus to $s$,
- the number of data communications required to multiply $M$ by a vector is independent of the number of processors $k^2$ while
- the amount of data transmitted per one processor is a decreasing function of the number of processors.

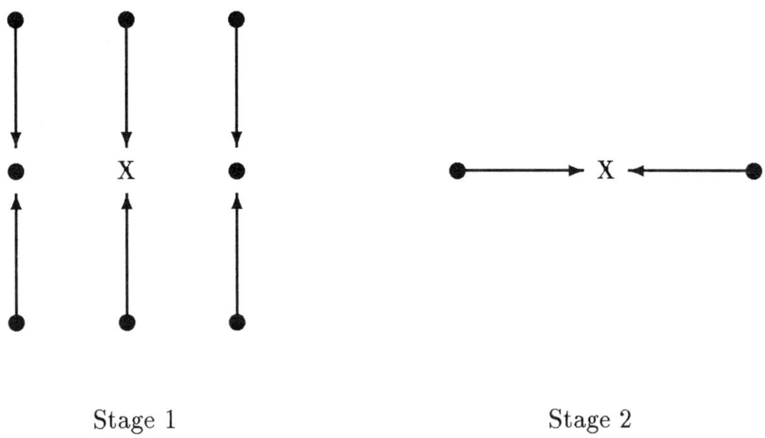


FIG. 4.2. *Scheme of data transmittings required to multiply a block row of $A_{ij}$ or $F_i$ by a vector on a 2D grid of processors with nearest neighbor connections*

The only serial work when performing an IBF-CG iteration is thus related to the inner product computations required by the CG algorithm and is obviously very small. As well known, each inner product can be computed and distributed among all processors in a number of communications proportional to the diameter of the communication network. Therefore, for a $k \times k$ grid of processors with nearest neighbor connections each inner product requires $2(k-1)$ data communications independently of the problem size.

Since the amount of parallel work for our implementation of an IBF-CG iteration is proportional to $s^3$ while communication overhead is proportional to $s$ we come to the important conclusion that for our implementation scheme the amount of overhead in exploiting parallelism together with the amount of serial work grow much slower than the amount of parallel work when the problem size increases. Therefore, a larger number of processors can be efficiently utilized when solving larger problems.

**5. Numerical results.** In this section we present convergence and complexity results for IBF-CG iterations with two IBF preconditioners in inverse free form applied to two test problems resulting from the first order finite element discretizations of the 3D equilibrium equations for a linear elastic orthotropic body (see [5]) with the Poisson ratios $\nu_{xy} = 0.35$, $\nu_{yz} = 0.37$, and $\nu_{xz} = 0.39$ on nonuniform curvilinear topologically cubic $20 \times 20 \times 20$ and $28 \times 28 \times 28$ meshes. The corrresponding SPD coefficient matrices $A_1$ and $A_2$ are of order $N(A_1) = 22800$ and $N(A_2) = 63504$ and have $NZ(A_1) = 1645924$ and $NZ(A_2) = 4741732$ nonzero entries, respectively.

Our numerical experiments aim at analyzing the dependence of the convergence rate and of the serial complexity of the IBF-CG algorithms on the sparsity pattern of the factorized sparse approximate inverses to pivot blocks. As a reference point we use the performance of the ICCG(0) algorithm which is commonly accepted as one of the best serial algorithms. It should be however noted that since $A$ is off-diagonally dominant we are forced to construct the incomplete Cholesky decomposition not for $A$ but for a shifted matrix $\widetilde{A} = \alpha I + A$ in order to preserve positive definiteness.

Two strategies for selecting the sparsity pattern when constructing factorized sparse approximate inverses to pivot blocks giving rise to two different inverse free IBF preconditionings are considered. The first one consists in prescribing for the lower triangular factor of the factorized sparse approximate inverse the sparsity pattern of the lower triangular part of the corresponding diagonal block of $A$. The second strategy takes explicitly into account the finite element origination of the matrix. Consider a uniform partitioning of the 2D $s \times s$ mesh into non-overlapping elements as shown in Fig. 3. Let all the nodes of a reference superelement be connected with all the nodes of the $\leq 9$ superelements adjacent to it and take as the sparsity pattern of the lower triangular factor of the factorized sparse approximate inverse that of the lower triangular part of the connectivity matrix of the constructed graph. Here the nodes of a plane $s \times s$ mesh are assumed to be numbered according to the natural order of non-overlapping elements and the natural order of nodes within elements.

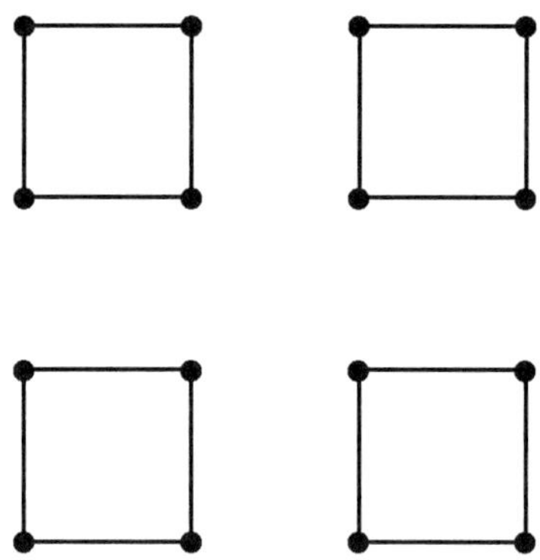

FIG. 5.1. *A uniform partitioning of a 4 by 4 plane mesh into non-overlapping elements*

Arithmetic costs for constructing the lower triangular factor of a factorized sparse approximate inverse are determined by the orders of linear systems to be solved to compute nonzero entries in individual rows of the former. Taking into account that the original PDE problem consists of three equations the maximum order of such systems does not exceed 15 for the first strategy and 60 for the second strategy independently of the size of the original mesh and is rather small relatively to the order of the original matrix. Note that nonzero entries of respectively 3 and 12 rows for the first and the second strategies can be computed by solving linear systems with the same coefficient matrix which enables one to save arithmetic work and introduce another type of parallelism into the preconditioner construction process.

*Remark* 5.1. Second strategy for selecting the sparsity pattern of a factorized sparse approximate inverse can easily be extended to the case of a non necessarily uniform partitionning of a 2D finite element mesh into non-overlapping superelements. Increasing in this way the number of nonzeros in the sparse approximate inverses to pivot blocks is expected to improve the preconditioning quality of the resulting IBF preconditioner but makes the construction of the latter and especially the computation of the products $A_{ii-1}F_{i-1}A_{i-1i}$ more expensive.

Tables 1 and 2 present convergence, complexity, and spectral characteristics for the ICCG(0) and two IBF-CG algorithms when applied to solving linear systems with the coefficient matrices $A_1$ and $A_2$ respectively and use the following notation:

*Iter* denotes the number of iterations required to satisfy the stopping criterion

$$||w_k||_2/||w_0||_2 < 10^{-11},$$

where $w_k$ is the pseudoresidual at the $k$'th iteration;

$AC_{Iter}$ shows the serial arithmetic costs of the required PCG iterations measured by the cost equivalent number of unpreconditioned CG iterations;

$AC_P$ stands for the serial arithmetic costs for the preconditioner construction also measured by the cost equivalent number of unpreconditioned CG iterations;

$\lambda_{\min}$ and $\lambda_{\max}$ denote respectively the smallest and the largest Ritz values for the preconditioned matrix while cond $= \lambda_{\max}/\lambda_{\min}$.

$IBF_{Stand}$ denotes the inverse free IBF preconditioning implemented using the straightforward computation of the product of the IBF preconditioned matrix times a vector, while

$IBF_{Opt}$ stands for the same preconditioning but implemented using formulae (4.7) – (4.9) to multiply the preconditioned matrix by a vector;

finally, *I* or *II* in the round brackets means respectively the first or the second strategy for selecting the sparsity pattern of factorized sparse approximate inverses to pivot blocks when constructing an IBF preconditioner.

Analyzing data in Tables 1 and 2 we come to the following conclusions.

- As expected, increasing the number of nonzeros in the IBF preconditioner when passing from strategy *I* to strategy *II* leads to a a quite considerable convergence acceleration. Moreover, for the second strategy of selecting the sparsity pattern of approximate inverses to pivot blocks the resulting convergence rate is much higher than that of the ICCG(0).

- Total serial arithmetic costs of the IBF-CG algorithm can be substantially decreased by using denser approximate inverses. Furthermore the serial costs of the iteration stage for the IBF-CG method with the second strategy of sparsity pattern selection and efficient computation using (4.7) – (4.9) of matrix-vector products are noticeably smaller than those for ICCG(0).

TABLE 5.1
*Convergence, complexity, and spectral characteristics for the ICCG(0) and IBF-CG algorithms for the Test Matrix of order N=22800*

| Preconditioning | Iter | $AC_{Iter}$ | $AC_P$ | $\lambda_{\min}$ | $\lambda_{\max}$ | cond |
|---|---|---|---|---|---|---|
| IC(0) | 389 | 803.23 | 5.4 | .713e-3 | 1.175 | .165e+4 |
| $IBF_{stand}(I)$ | 774 | 1888.56 | 49.49 | .139e-2 | 7.382 | .531e+4 |
| $IBF_{opt}(I)$ | 774 | 1327.08 | 49.49 | .139e-2 | 7.382 | .531e+4 |
| $IBF_{stand}(II)$ | 250 | 853.50 | 457.56 | .545e-2 | 7.792 | .143e+4 |
| $IBF_{opt}(II)$ | 250 | 853.50 | 457.56 | .545e-2 | 7.792 | .143e+4 |

TABLE 5.2
*Convergence, complexity, and spectral characteristics for the ICCG(0) and IBF-CG algorithms for the Test Matrix of order N=63504*

| Preconditioning | Iter | $AC_{Iter}$ | $AC_P$ | $\lambda_{\min}$ | $\lambda_{\max}$ | cond |
|---|---|---|---|---|---|---|
| IC(0) | 714 | 1481.31 | 5.4 | .241e-3 | 1.235 | .512e+4 |
| $IBF_{stand}(I)$ | 1244 | 3035.36 | 51.22 | .620e-3 | 11.57 | .187e+5 |
| $IBF_{opt}(I)$ | 1244 | 2276.52 | 51.22 | .620e-3 | 11.57 | .187e+5 |
| $IBF_{stand}(II)$ | 476 | 1623.16 | 505.90 | .139e-2 | 8.501 | .611e+4 |
| $IBF_{opt}(II)$ | 476 | 1308.32 | 505.90 | .139e-2 | 8.501 | .611e+4 |

*Remark* 5.2. When performing the numerical experiments reported no scaling ensuring condition (3.4) was utilized.

Taking into account (i) the natural parallelism of constructing factorized sparse approximate inverses and (ii) parallel properties of inverse free IBF preconditioned CG iterations we see that IBF-CG algorithms provide a promising approach to solution of three dimensional problems on parallel computers with moderate numbers of processors.

**6. Conclusion.** We have presented two methods for constructing incomplete block factorization preconditioners for SPD block tridiagonal matrices. The inverse free incomplete block factorizations for 3D finite element problems were shown to possess an efficient parallel implementation on a two-dimensional grid of processors. An important property of this implementation is that when the problem size grows the amount of both serial work and communication overhead increases much slower than the amount of parallel work. On the other hand, as demonstrated by numerical results, with an appropriate choice of block tridiagonal partitioning and sparse approximations to inverses of pivot blocks IBF-CG algorithms converge much faster than the ICCG(0) method which is one of the best serial algorithms.

**7. Acknowledgement.** The authors are grateful to A. Nikishin for his valuable help in running numerical experiments.

## REFERENCES

[1] O. Axelsson, *Incomplete block matrix factorizations preconditioning methods. The ultimate answer?* J. Comput. Appl. Math. 12&13, 3–18, 1985.

[2] O. Axelsson and L. Kolotilina, *Diagonally compensated reduction and related preconditioning methods*, Rept. 9117, Dept. of Math., Catholic University, Nijmegen, Aug. 1991, to appear in J. of Numer. Linear Algebra with Appl.

[3] O. Axelsson and B. Polman, *On approximate factorization methods for block matrices suitable for vector and parallel processors*, Linear Algebra Appl. 77, 3–26, 1986.

[4] P. Concus, G.H. Golub, and G. Meurant, *Block preconditioning for the conjugate gradient method*, SIAM J. Sci. Stat. Comput. 6, 220–252, 1985.

[5] I. Kaporin, L. Kolotilina, and A. Yeremin, *Block SSOR preconditionings for high 3D order FE Systems II: Incomplete BSSOR preconditionings*, Linear Algebra Appl. 154–156, 647–674, 1991.

[6] L. Kolotilina and B. Polman, *On incomplete block factorization methods of generalized SSOR type for H-matrices*, Rept. 9120, Dept. of Math., Catholic University of Nijmegen, Sept. 1991, to appear in Linear Algebra and Appl..

[7] L. Kolotilina and A. Yeremin, *Factorized sparse approximate inverse preconditionings I: Theory*, to appear in SIAM J. on Matrix Analysis Appl. vol. 14, Jan. 1993.

[8] L.Kolotilina and A.Yeremin, *Block SSOR preconditionings for high order 3D FE Systems*, BIT, vol. 29, No 6, 1989.

[9] A. Yeremin and L. Kolotilina, *Incomplete block factorization methods for complex-structure matrices* (in Russian), Zap. Nauch. Semin. LOMI 159, 5–22, 1987; Engl. translation: J. Soviet Math. 47, 2821–2834, 1989.

# HOW FAST CAN ITERATIVE METHODS BE?

OLAVI NEVANLINNA*

**1. Background.** During the last two years I have been trying to formulate answers to this question: how fast can iterative methods be for linear problems. Together with Timo Eirola we constructed about five years ago a family of Krylov subspace methods and while testing them later, we noticed that the speed (measured in terms of matrix-vector products) was quite similar over different Krylov subspace methods, compared with the sensitivity due to different preconditioning strategies. Thus: what are the properties of the matrix or operator which results after preconditioning that allow fast acceleration?

I have given lectures on this topic at Helsinki University of Technology in the fall 1990 and at ETH in Zurich in 1992. I identify three different mechanism in the convergence of a Krylov subspace method which correspond to *i) sublinear, ii) linear, and iii) superlinear* convergence. The material on sublinear convergence is separately available as a report [4] from spring 1991. The whole set of lecture notes [5] is planned to be out by the end of 1992.

In this paper I try to overview some of the ideas and the type of mathematics used. I want to stress that my work does not intend to be a survey of what is known. The work is rather an attempt to widen the vocabulary in this field.

**2. Why Banach spaces?** With the software of today one typically first discretizes the original mathematical problem and then solves the resulting linear algebra problem with tools from numerical linear algebra.

In the future we shall see more and more of the following: we have good, robust and fast that is, software to treat some basic problems and we use such software to approximately invert our original problem. This naturally leads to iterative techniques and is not confined to fixed-dimensional linear algebra (domain decomposition and waveform relaxation are examples of this). In this way one preconditions the original problem directly and not the linear algebra problem at the end of the reduction chain.

One of the main points of Krylov subspace methods is that they only assume a piece of software to be given which creates the matrix-vector products at a given vector. In this respect nothing changes if the piece of software creates outputs into a function space. I chose to mainly work in Banach spaces rather than in Hilbert spaces because my main motivation was to look at nonsymmetric problems: the temptation of relying to spectral decomposition of normal operators is then excluded. (This choice is

* Helsinki University of Technology, Institute of Mathematics, SF-02150 Espoo, Finland.

for analysis; it is another matter to implement the methods.)

The mathematics used in numerical linear algebra draws much, perhaps too much, from the spectral decomposition. This polarizes the landscape to "good" matrices and "badly nonnormal" matrices.

When excluding the spectral decomposition, we must assume some symbolic calculus for the sake of estimating polynomial operators. The natural and simplest is provided by the Cauchy integral:

$$\phi(L) = \frac{1}{2\pi i} \int_{\Gamma} \phi(\lambda)(\lambda - L)^{-1} d\lambda. \tag{2.1}$$

Here $L$ is a bounded operator in a Banach space, $(\lambda - L)^{-1}$ the resolvent operator, $\phi$ analytic in a neighborhood of the spectrum $\sigma(L)$ (the neighborhood is assumed to have only finitely many components) and $\Gamma$ surrounds the spectrum inside this neighborhood.

When we want to estimate $||\phi(L)||$ effectively, *growth properties* of the resolvent operator near the spectrum are important, and thus the primary focus is in thinking of $(\lambda - L)^{-1}$ as an analytic operator-valued function in $\lambda$, defined outside the spectrum. *Pseudospectrum* has been proposed as an intermediary concept between spectrum and the resolvent. In practice this corresponds to restricting oneself to integrate along level curves of $||(\lambda - L)^{-1}||$.

For much of the discussion it is irrelevant whether the space is infinite or finite dimensional. However, trivially, in finite dimensional space the operators have a small (finite) spectrum and a good Krylov subspace method always terminates with finite number of steps. The point is, that one wants to stop much earlier than what the dimension promises, and therefore it is very useful to understand the asymptotic convergence properties of infinite dimensional problems, the more so as the matrices often approximate some operators in infinite dimensional spaces.

**3. SOR and CG.** In retrospect there are a couple of interesting questions, somewhat of sociological nature, concerning the history of (preconditioned) iterative methods (or semi-iterative, whatever language one prefers). The first obvious one is: *why the preconditioned conjugate gradient method did not appear immediately right from the beginning, forthy years ago, in place of SOR?* The other one is: *why our understanding on iterating nonsymmetric problems is still so incomplete?* It is important for the progress of science to pose questions of these nature. I dont try to answer these questions, but I shall explain in some length the one answer which motivates the approach I am taking.

In a sense, both SOR and CG were given to the scientific community in a too fine, final form. It was possible to explain their behavior by quite elementary mathematics, in particular avoiding complex analysis. Numerical analysis students learned how, say for the discretized Laplacian, the spectral radius of SOR, $\rho(G_\omega)$ behaves as the relaxation parameter $\omega$

increases, but we (or I at least) were not told how the eigenvalues behave. Likewise Chebyshev polynomials were (and are) used to derive bounds for the error in CG. Again it gives the correct answer but fails to uncover the "reason" (a student cannot know why Chebyshev polynomials decay the way they do; a numerical analyst might look at a book on Chebyshev polynomials and find a calculation using real analysis tools as the problems sits entirely on an interval on the real line).

So, let me formulate a specific question: *why SOR and CG are essentially equally fast for discretized Laplacians* (by CG I mean that you accelerate the Jacobi iteration), *what is the intuitive heuristic reason, why can you not do faster.*

Answer: if you iterate an operator which has a circular spectrum, centered at the origin, then the best you can do is just the basic iteration, no acceleration is possible. Both SOR and CG *essentially* transform the original situation back to this basic iteration.

Consider first SOR where the phenomena is exactly true (and well known). For $\omega = 1$ all eigenvalues are real and, if we denote by $G$ the iteration operator, then

$$\rho(G) = 1 - O(\epsilon^2)$$

where $\epsilon$ is a discretization parameter. When $\omega$ increases then the eigenvalues collide, form complex pairs which at optimal $\omega$ are on a circle of minimal radius

$$\rho(G_\omega) = 1 - O(\epsilon).$$

Thus, the overrelaxation massages the operator in such a way that the spectrum gets mapped from an interval into a circle. Now, CG almost does the same. We need to identify what is it that almost exists and almosts gets raised to power in the iteration. Let $J := D^{-1}(L+U)$ be the preconditioned operator. Then we again have the eigenvalues of $J$ on an interval with

$$\rho(J) = 1 - O(\epsilon^2).$$

Now CG does not change the operator itself as it works within the subspace spanned by the images of powers of $J$. However, we can imagine the forming of a new operator $\phi(J)$ where $\phi$ would be an analytic function mapping the interval where the spectrum lies onto a small circle. Then the iterates would decay with the rate given by

$$\rho(\phi(J)).$$

By the spectral mapping theorem we would have

$$\rho(\phi(J)) = \sup_{\sigma(J)} |\phi(\lambda)|. \tag{3.1}$$

The right hand side of (3.1) does give the correct linear decay rate, but the operator $\phi(J)$ does not exist. The problem is that a function mapping the interval onto a circle is not analytic in any neighborhood of the interval and (2.1) cannot be used to define $\phi(J)$. If we instead of asking for an analytic $\phi$ in a neighborhood of the spectrum, we ask for a $\phi$ which is analytic *outside* of the spectrum (or interval) then the outside of the interval can be mapped onto outside of a disc, centered at the origin. So, let $\phi$ be this conformal map (growing linearly as $\lambda \to \infty$, and satisfying $\phi(1) = 1$). Consider the integral on the right hand side of (2.1). It is well defined. More generally, consider

$$A_k(J) := \frac{1}{2\pi i} \int_\Gamma \phi(\lambda)^k (\lambda - J)^{-1} d\lambda \tag{3.2}$$

where $\Gamma$ surrounds the spectrum. In particular we obtain that for some $C = C(\Gamma)$

$$||A_k(J)|| \le C\eta^k \tag{3.3}$$

where $\eta = \sup_\Gamma |\phi|$. The integral is independent of $\Gamma$ as long as it surrounds the spectrum. Next, observe that $A_k(J)$ is a *polynomial* in $J$. In fact, define the (Faber) polynomials $F_k$ by requiring

$$\phi(\lambda)^k = F_k(\lambda) + O(1/\lambda) \text{ as } \lambda \to \infty. \tag{3.4}$$

Then, integrating over a very large circle,

$$A_k(J) - F_k(J) = \frac{1}{2\pi i} \int_\Gamma [\phi(\lambda)^k - F_k(\lambda)](\lambda - J)^{-1} d\lambda$$

and (3.4) imply $A_k(J) = F_k(J)$.

CG works with minimizing residuals $r^k$ so that necessarily

$$||r^k|| \le ||p_k(J)r^0|| \tag{3.5}$$

holds for all polynomials $p_k$, normalized to satisfy $p_k(1) = 1$ (we are looking for fixed points, not for zeros) and of degree $\le k$. Since it is easy to see that $F_k(1) \sim \phi(1)^k = 1$ we obtained the desired bound by combining (3.3) and (3.5).

It is now clear why the acceleration is from $\rho(J) = 1 - O(\epsilon^2)$ to $1 - O(\epsilon)$: the conformal map $\phi$ maps like a square root at the end of the interval, near the normalizing point. Thus the distance $O(\epsilon^2)$ gets increased dramatically to $O(\epsilon)$.

(Eventually, the sparseness of the spectrum of a discretized Laplacian shows up and CG behaves superlinearly; thus tho say that the linear rate above is "correct", is a complicated statement which would require a longer discussion.)

Observe that in the discussion above we did not need any assumption on $J$ other than its spectrum is inside an interval. The discussion generalizes to any operator whose spectrum can be included effectively inside a set whose complement is simply connected on the extended plane.

This, however, is often *not* the case. We simply have to give up the idea of "almost an operator" whose spectral radius we would look at. What we can still save is the notion of the linear speed given by a "radius." In the same way as the resolvent operator codes the spectral radius as the radius outside which the Laurent series

$$(\lambda - L)^{-1} = \sum L^k \lambda^{-1-k}$$

converges, I associte to a given polynomial sequence $\{p_k\}$ the generating function

$$G(\zeta, L) := \sum_{k=0}^{\infty} p_k(L)\zeta^{-1-k} \tag{3.6}$$

which then codes the obtainable speed: the smaller the radius outside which this is analytic, the faster the (linear) rate. If this is analytic everywhere outisde the origin we have superlinear convergence. The main idea is thus to use the machinery of complex analysis on such generating functions.

**4. Some notation.** Let the original problem be $Ax = b$. If we precondition with $H$ then we end up with a fixed point problem of the form

$$x = Lx + g \tag{4.1}$$

where $L := 1 - AH$ (or $L = 1 - HA$ or like) is the error operator in the approximate inversion of $A$. Writing the resulting equation into a fixed point form exhibites the *semigroup-like* dynamical properties of the iteration in a natural way (see the discussion on CG in the previous section: $p_k(L) \sim$" $\phi(L)^k$j " I shall assume that $L$ is a bounded linear operator in a (finite or infinite dimensional) Banach space, not necessarily such that $1 - L$ is nonsingular; if it is singular then additionally $g$ is in the range of $1 - L$ so that (4.1) has a solution. Further, $p_k$ denotes any polynomial of degree $\leq k$ such that $p_k(1) = 1$. To each sequence there is associated a generating function $G(\zeta, L)$ by (3.6) which we primarily consider as a operator valued analytic function in $\zeta$, for a fixed $L$.

The spectrum of $L$, $\sigma(L)$ is a compact set in the plane; and any compact set is a spectrum of some operator. The complement of $\sigma(L)$ need not be connected, but in any case it has exactly one unbounded component, which I denote by $G_\infty(L)$. Other components of the complement of the spectrum play no role in our topic.

**5. The lifespan of an iteration.** It is customary to plot the norms of residuals (or errors) into a logarithmic scale, against the iteration index.

We then obtain a more or less straight line. To my eye, plotting the ratio of residuals, $||r^k||/||r^{k-1}||$ reveals more of the particular method and problem. Let us call that the *lifespan* of the iteration. The plot can be made to look like almost any curve by tailor making a strange operator and initial residual, but it often looks like a tea cup, upside down. In the lifespan I expect to see the following phases: in the early sweeps the convergence is very rapid but slows down, this is the sublinear behavior. The convergence then settles down to a roughly constant linear rate: the plot looks horizontal. Towards the end new speed may be picked up again, corresponding to the superlinear behavior. Finally it terminates either by forced truncation or by hitting the goal exactly.

In practice all phases need not be identifiable, nor need they appear only once and in this order. For example, a SOR iteration does not reach the superlinear phase, but before settling down to a linear speed it may converge faster, and this would correspond to a sublinear, not superlinear behavior.

Heuristics:

i) *sublinear phase.* Initially, when starting the iteration there is no information to adapt to. The method does even not "know" whether the problem is nonsingular or not (I call it singular if $1 \in \sigma(L)$). Thus we conclude that the speed can also be described with terms not depending on the singularity. In particular, nothing like the condition number $||1 - L||||(1 - L)^{-1}||$ plays initially any important role. Imagine looking the spectrum initially from far away. Typically most of the spectrum would be to left from 1, maybe within a sectorial set but you dont see whether the problem is singular or not. A "good" Krylov subspace method starts to "examine" the spectrum from the far left, working its way towards the normalizing point 1 following the "outer boundary" of the spectrum. This can be visualized by plotting the evolution of the zeros of the residual polynomials created. Eventually these zeros come close to 1 and the method "learns" that the problem is nonsingular (assuming it was). Then also the convergence behavior changes from sublinear to linear (or directly to superlinear) phase.

Sublinear convergence is analogous to the behavior of analytic semigroups. Language in ergodicity theory also appears naturally.

ii) *linear phase.* The linear phase exists as long as the "outer boundary" of the spectrum has not been completely killed by the zeros of residual polynomials. For a matrix, by outer boundary I mean a rough curve drawn thru the extreme eigenvalues; it may consist of a "small" number of different continua. The linear rate is simply given by the Green's function for the outside of the outer boundary. If it consists of one component only, then the Green's function can be given by the conformal map discussed in the connection of CG above.

When the spectrum at the outer boundary (or on a component of it) has been killed by the zeros the remaining effective spectrum looks smaller and the process starts to accelerate. This means that we move to superlinear phase.

iii) *superlinear phase.* The superlinear phase goes together with clustering of the spectrum. Mathematically, any point other than 1 is essentially equally natural as a cluster point, but in case of just one cluster it is natural to think that the cluster is centered at the origin (which we could arrange by translating and scaling the operator without changing the Krylov-information). Then the operator has a lot of small eigenvalues and I have bounds for the superlinear behavior in terms of the decay of the singular values.

The iteration eventually goes into superlinear phase if and only if the capacity of the spectrum vanishes (and the problem is nonsingular). For example, countable sets are of zero capacity and thus superlinear convergence is there for example whenever the operator is compact. In finite dimensional cases the superlinear behavior is formally always there, say for GMRES; if the singular values decay too slowly the finite termination property takes over.

To describe the superlinear convergence I use terminology and tools from the theory of entire functions. The generating function $G(\zeta, L)$ is then analytic outside origin and is thus entire in $1/\zeta$.

Linear RC-networks give us an example where all three phases can persist. In fact, in waveform relaxation the convergence can be asymptotically sublinear, linear or superlinear, depending on the decopling of the network. To learn about this, look e.g. at the references in [2].

iv) *finite termination.* An algorithm with the finite termination property (terminates at the exact solution after finite number of (exact arithmetic) steps) exists for a particular $L$ if and *only if* there is an annihilating polynomial for $L$: $p_n(L) = 0$. This is of course trivial in finite dimensional spaces but nontrivial in infinite dimensions. As this is very simple to state, I formulate it here:

THEOREM A. *Given a bounded linear operator $L$ in a Banach space the following holds. Either there exists a vector $x$ such that for all normalized polynomials $p_k$*

$$p_k(L)x \neq 0$$

*or there exists a normalized $p_n$ such that*

$$p_n(L) = 0.$$

(Theorem 5.4.4 in [5]; This is a consequence of a result of I. Kaplansky)

In general Banach spaces operators with finite rank do have the finite termination property. It is thus natural to expect that if we can approximate a given operator well with low-srank operators then we have fast convergence. This is one natural way to think why fast decay of singular values implies fast convergence. However, operators like multiplication by a scalar does have the finite termination property but it is extremely poorly approximated with finite rank operators. In general, if there exists a polynomial $Q$ (of minimal degree $n$) such that $Q(L) = 0$ then we say that $L$ is an *algebraic operator of degree* $n$. Analogously to singular values $\{\sigma_n\}$ I define *approximation numbers* by algebraic operators:

$$\alpha_n := \inf \|L - A_n\| \text{ for } n = 1, 2, \ldots$$

where the infimum is over all algebraic operators $A_n$ of degree $\leq n$. We have always

$$\alpha_n \leq \sigma_n \text{ for } n = 1, 2, \ldots$$

They can decay much faster than the singular values, but if they decay fast enough (in fact superlinearly) then "good" Krylov methods achieve superlinear speed.

**6. Is the spectrum all I care about?** The reader might wonder why I talk about the spectrum as if everything would be explained by knowing it. First of all, all quantitative results except the asymptotic linear speed do depend on other things like the growth of the resolvent and these of course are included in the assumptions. However, cases where the spectrum gives truely a wrong picture of the convergence are pathological and fall typically into one of the following categories:

1. The spectrum is much "smaller" than the set where the resolvent is large

2. The initial residual is not "generic" for the operator

If the spectrum is unnaturally small, then there is not much "reason" for a linear phase; the eventual beginning of the superlinear phase can be estimated e.g. in terms of the singular values.

The second case has been looked at by Anne Greenbaum [1]. She has studied the set of matrices which generate (starting from a given residual) the same Krylov subspaces and shown that it always contains e.g. a unitary matrix. *Dilation theory* can be used to give another, different view of this: any bounded operator $A$ in a Hilbert space (scaled down to have norm at most 1) can be viewed as a restriction of a unitary operator, say $U$ in a larger Hilbert space in such a way that for all $x$ in the original space

$$A^k x = PU^k x, \text{ for all } k = 0, 1, \ldots$$

where $P$ is the orthogonal projection from the large space to the original space. Thus we can view the Krylov subspace as a restriction of one created by a unitary operator, but then the initial residual is sitting in a "small" original space.

So, is the spectrum all I care about? In a nutshell the answer is: Of course not, but the very fact that spectrum does not tell it all should not direct us to search for another *single concept* which would tell it all.

**7. Sublinear phase.** Let $\hat{p}_k(L)$ be a smallest polynomial for $L$ in the sense that

$$\|\hat{p}_k(L)\| \leq \|p_k(L)\| \tag{7.1}$$

for all $p_k$ of degree $\leq k$ satisfying $p_k(1) = 1$. Then with $p_{n+m} := \hat{p}_n\hat{p}_m$

$$\|\hat{p}_{n+m}(L)\| \leq \|\hat{p}_n(L)\| \; \|\hat{p}_m(L)\| \tag{7.2}$$

Applying $p_k(\lambda) \equiv 1$ we see that $\|\hat{p}_k(L)\| \leq 1$. This does not leave much freedom for *sublinear* convergence of the operator norms as either $\|p_k(L)\| \equiv 1$ or the convergence is *eventually at least linear.*

We say that a sequence $\{a_k\}$ of positive reals *decays sublinearly with exponent* $\beta$ if

$$\beta := \liminf_{k\to\infty} \frac{\log(1/a_k)}{\log k}$$

is finite and then we can write roughly

$$a_k \sim \frac{C}{k^\beta}.$$

Such a sequence cannot satisfy an inequality of the form (7.2). As an analogy, think about linear operator semigroups; let $A$ generate a contraction semigroup $e^{tA}$ . Say, for the wave equation we do not obtain any decay while for the heat equation the *time derivative* decays. The Laplacian generates an *analytic* semigroup and for all analytic semigroups we have

$$\|Ae^{tA}\| \leq \frac{C}{t}, \text{ for } t > 0$$

( with some $C$; if $C < 1/e$ then $A$ is bounded). Contraction semigroups cannot decay faster than exponentially (which is the analogy of linear convergence). The sequence $\{p_k(L)\}$ on the other hand can decay superlinearly (and even terminate to zero in finite time) so the possible parallelism is only onesided which is clear as we replace $e^{(t+s)A} = e^{tA}e^{sA}$ by the more liberal inequality (7.2) corresponding to $\|e^{(t+s)A}\| \leq \|e^{tA}\| \; \|e^{sA}\|$.

THEOREM B. *Assume that $G_\infty(L)$ contains an open sector $S_\alpha$ with top at 1 and with opening $\alpha$. Assume further that the resolvent satisfies*

$$\|(\lambda - L)^{-1}\| \leq \frac{Const}{|\lambda - 1|}.$$

*Then for every $\epsilon > 0$ there exists polynomials $\{p_k\}$ satisfying*

$$\|p_k(L)(1 - L)\| \leq \frac{C}{(k+1)^\kappa} \tag{7.3}$$

*Here $C$ is a constant independent of $k$ and $\kappa := \frac{\alpha}{\pi} - \epsilon$.*

(Theorem 4.6.1, in [4])

For example, consider CG in the case of the discretized Laplacian. The spectrum sits on an interval, and (by symmetry of the matrix) the resolvent satisfies the assumptions above in any sector of opening less than $2\pi$. Thus we can take $\kappa := 2 - \epsilon$. Likewise for typical *indefinite* problems we obtain $\kappa := 1 - \epsilon$.

In the inequality (7.3) the "difference" $1 - L$ can be replaced by any fractional power $(1-L)^s$ with $s > 0$. Then the constant $C$ depends on $s$ as well. The dependency is of the following nature: $C(s) = O(1/s)$ as $s \to 0$ while $C(s) = O(s^{\kappa s})$ as $s$ grows. The inequality is then

$$\|p_k(L)(1-L)^s\| \le \frac{C(s)}{(k+1)^{\kappa s}}. \tag{7.4}$$

Thus the exponent $\beta$ is a product of two terms, $\kappa$ coming from the largest sectorial set into which $G_\infty(L)$ (the unbounded component of the resolvent set) opens, when looked from 1, while $s$ codes the distribution of the initial error. In fact, if the initial residual $r^0$ can be written as

$$r^0 = (1-L)^{s+1}a \tag{7.5}$$

with some vector $a$, then the error satisfies

$$e^k = p_k(L)(1-L)^s a. \tag{7.6}$$

Notice that for small values of $k$ the bound promises a *fast decay which however slows down.*

If $1-L$ is nonsingular we can write $r^0$ in the form of (7.5) for any $s > 0$ and optimize the bound. To that end assume that

$$\|(1-L)^{-n}\| \le C_1 R^n$$

for some $C_1$ and (typically large) $R$. Let us assume an inequality of the form (7.4) in the quantitative form

$$\|p_k(L)(1-L)^n\| \le \left(\frac{C_0 n}{k+1}\right)^{\kappa n}.$$

Then for large enough $k$ we choose $n \sim \frac{k}{C_0 e R^{1/\kappa}}$ and obtain

$$\|e^k\| \le C_1 e^\kappa \eta^k \|e^0\| \tag{7.7}$$

where

$$\eta := e^{-\kappa / C_0 e R^{1/\kappa}}.$$

Notice that this is simply a bound for linear decay. Applying (7.7) to our standard example of CG and the discretized Laplacian we take $\kappa = 2 - \epsilon$ and obtain $\eta = 1 - O(\epsilon)$ which is of the correct order of magnitude.

**8. Linear phase.** Let us put

$$\eta(L) := \inf \|p_k(L)\|^{1/k} \tag{8.1}$$

where the infimum is over all integers $k \geq 0$ and over all polynomials of degree $k$ satisfying the normalization $p_k(1) = 1$. The error can be reduced asymptotically at this rate and we call it the *optimal reduction factor* [3].

The result giving the value of $\eta(L)$ takes a simple form.

THEOREM C. *We have*

*i)* $\eta(L) < 1$ *if and only if* $1 \in G_\infty(L)$

*ii)* $\eta(L) = 0$ *if additionally* $cap(\sigma(L)) = 0$.

*For* $0 < \eta(L) < 1$ *we have*

$$\eta(L) = e^{-g(1)}$$

*where* $g$ *is the Green's function for* $G_\infty(L)$.

*As a function of* $L$ $\eta(L)$ *is upper semicontinuous (but in general not continuous).*

(Theorem 3.4.9 in [5]) Here the capacity is the logarithmic capacity on the plane; for a compact set of positive capacity there exists a Green's function which is defined in the outer component, grows logarithmically at infinity and vanishes when the boundary is approached, except possibly on a small polar set. No regularity of the boundary is needed. If $G_\infty(L)$ is simply connected (in the extended plane) then the conformal map $\phi$ mapping it to outside of a disc gives the Green's function: $g = \log|\phi|$ (if $\phi$ is normalized suitably). The proof of the theorem is quite simple: first one shows, essentially with the help of the spectral radius formula that $\eta(L)$ depends only on the spectrum. Then the problem is reduced to approximation and potential theory: you have to approximate 0 by polynomials uniformly in a compact set.

There are a lot of results related to this statement in the literature. Often they are, however, formulated for the matrix case: as the capacity of the spectrum of a matrix is always 0 it is usually assumed that one includes the spectrum into a larger set with a smooth boundary and considers the Green's function or the conformal map associated with this larger set instead.

In section 5 I said that the linear speed roughly lasts as long as the outer boundary has not been killed by the zeros of the polynomials $p_k$. If the capacity of the spectrum is positive, then this never happens. For circularly symmetric spectra one can show that the outer spectrum dominates exactly as long as the iteration count is smaller than the number of points in the outer boundary.

**9. Superlinear phase.** Let $\{\hat{p}_k\}$ be a sequence of smallest polynomials in the sense of (7.1). Put

$$G(\zeta, L) := \sum_{j=0}^{\infty} \hat{p}_k(L)\zeta^{-1-k}. \tag{9.1}$$

From the definition of $\eta(L)$ it follows (as in the scalar case) that $G(\zeta, L)$ is analytic in $\zeta$ outside of a disc of (exact) radius $\eta(L)$. In particular, when $\text{cap}(\sigma(L)) = 0$ so that $\eta(L) = 0$, then $G$ is analytic in $\zeta$ for $\zeta \neq 0$. In other words, $G$ is entire in the variable $1/\zeta$.

One says that (a scalar function) $f(z)$ is entire of *order* $\omega$ and of *type* $\tau$ if (roughly)

$$|f(z)| \sim e^{\tau |z|^\omega} \text{ as } z \to \infty.$$

If $f(z) = \sum a_j z^j$ then equivalently

$$|a_j| \sim (\frac{\tau e \omega}{j})^{j/\omega} \quad \text{as } j \to \infty.$$

The same equivalence result holds also in our operator valued case: the growth of the generating functions as $\zeta \to 0$ tells how fast the coefficient polynomials decay. This gives us a language to be used. However, for numerical purposes we would rather work with bounds than with asymptotic concepts. A natural bound takes the following form

$$\|p_{k+N}(L)\| \leq C(\frac{\tau e \omega}{k})^{k/\omega} \text{ for } k > 0 \tag{9.2}$$

We could say that $\{p_k(L)\}$ *admits a superlinear bound of order* $\omega$, *type* $\tau$, *with constant* $C$ and with a (possible) *delay* $N \geq 0$.

If the spectrum clusters at several places, the centers can be mapped to 0 with a polynomial. This reduces us to study the case where the spectrum accumulates at the origin. I shall formulate one result in full detail.

Let $L$ be compact in a separable Hilbert space (so that a matrix in a finite dimensional space will do) and let $\{\sigma_n\}$ be its singular values. Assume

$$\sum_n \sigma_n^\omega < \infty \tag{9.3}$$

holds for some fixed $\omega$ such that $0 < \omega \leq 1$. Let $\{\lambda_j\} = \sigma(L)$ be its eigenvalues (in decreasing order of absolute values) and put

$$C := \sup_{k \geq 1} \frac{1}{\prod_{j=1}^k |1 - \lambda_j|} \tag{9.4}$$

to denote the "conditioning" of the spectrum.

THEOREM D. *Under the assumptions above there are normalized polynomials* $p_k$ *such that*

$$\|p_k(L)\| \leq C(\frac{\tau e \omega}{k})^{k/\omega} \textit{ for } k > 0, \tag{9.5}$$

*where*

$$\tau \leq \frac{1+\omega}{\omega} \sum_j \sigma_j^\omega .$$

(Theorem 5.8.10 in [5].)

I have also shown that if there are $N$ large singular values but the rest are small, then a similar estimate with delay $N$ holds, where the type $\tau_N$ can be estimated in terms of the remaining singular and eigenvalues, excluding the $N$ largest ones.

In general, if the operator is compact then the spectrum is at most countable and in particular of capacity zero. Thus, with compact operators we always achieve superlinear convergence.

Let $\{\alpha_n\}$ be the approximation numbers by algebraic operators, see section 5. For a noncompact operator it is possible that $\alpha_n \to 0$ but $0 < \eta(L) < 1$. However, if the numbers decay fast enough, then we do have a superlinear convergence.

THEOREM E *If* $\alpha_n^{1/n} \to 0$ *as* $n \to \infty$ *and* $1 \notin \sigma(L)$, *then* $\eta(L) = 0$.

(This follows from the material of section 5.10 in [5]).

This "paper" is essentially a written version of a talk at IMA on March 3rd, 1992. I dedicate it to Gene Golub in the occation of his 60th birthday; he invited me to submit a paper.

## REFERENCES

[1] Anne Greenbaum, Lecture at IMA in February 1992.

[2] Ulla Miekkala and Olavi Nevanlinna: Quasinilpotency of the operators in Picard-Lindelöf iteration, Numer.Funct.Anal. Optimz. 13, 203-221, (1992).

[3] Olavi Nevanlinna: Linear acceleration of Picard-Lindelöf iteration, Numer. Math. 57, 147-156 (1990).

[4] Olavi Nevanlinna: Lectures on Iterative Methods: The Sublinear Case, Helsinki University of Technology, Mathematics Report C8, May 1991.

[5] Olavi Nevanlinna: Lecture Notes to be published by Birkhäuser.

# RATIONAL KRYLOV ALGORITHMS FOR NONSYMMETRIC EIGENVALUE PROBLEMS

AXEL RUHE *

**Abstract.** The Rational Krylov algorithm for the nonsymmetric matrix pencil eigenvalue problem is described. It is a generalization of the shifted and inverted Lanczos (or Arnoldi) algorithm, in which several shifts are used in one run. It computes an orthogonal basis and a small Hessenberg pencil. The eigensolution of the Hessenberg pencil, gives Ritz approximations to the solution of the original pencil.

Rational Krylov is the natural alternative when factorization of the matrix is not much more expensive than solution, as e. g. when a multigrid algorithm is used to solve systems.

In one variant several iterations with different shifts are started on the same starting vector. Such iterations can be performed in parallel, yielding a $p$ degree Krylov vector in one iteration on $p$ processors. An analogy to a method of experimentally verifying stability of aircraft structures is shown.

The algorithm is demonstrated on two applied test problems.

**AMS(MOS) subject classifications.** 65F15

**1. Introduction and summary.** In the present contribution, we develop an algorithm for computing eigenvalues of regular matrix pencils,

$$(A - \lambda B)x = 0 \ , \tag{1.1}$$

where the matrices, $A$ and $B$, are square but nonsymmetric, too big to be treated by similarity transformations, but not too big to be factorized,

$$A - \mu B = LU \ , \tag{1.2}$$

by sparse Gaussian elimination. We are interested in a subset of all the eigenvalues, most often those in a prescribed region of the complex plane as e. g. those closest to the origin.

Our aim is to continue the development of Krylov subspace algorithms, which started with Lanczos [10], see also [13,14,5,17] for further developments, and was continued for symmetric definite pencils (1.1) with shift and invert strategy (1.2) in [4,6]. The Arnoldi algorithm [1,20] has been brought out as a simple minded alternative to the two sided Lanczos algorithm for nonsymmetric pencils (1.1). Though it does not break down the way two sided Lanczos does, a two sided Arnoldi [18] is actually necessary for a convergence theory similar to [3,15,7] to hold.

All these algorithms work by computing bases of Krylov spaces. In the simplest case, $B = I$, the $j$th Krylov space is

$$K_j(A, x) = \operatorname{span}\{x, Ax, \ldots, A^{j-1}x\} \ . \tag{1.3}$$

* Chalmers University of Technology and the University of Göteborg, Department of Computer Sciences, S-41296 Göteborg, Sweden, (na.ruhe@na-net.ornl.gov). The work was partly supported by the Swedish National Board for Technical Development, STU grant 91-1210.

Any vector $v$ in the Krylov space (1.3) is a polynomial in the matrix $A$ acting on the starting vector $x$,

$$v = p_{j-1}(A)x \ .$$

In each step $j$, we add one to the degree of the polynomials. Orthogonal vectors correspond to orthogonal polynomials, and we can use a multitude of results from the theory of polynomial approximation to shed light on what happens in these Krylov based matrix iterations.

The shift and invert strategy brings rational functions into the picture, the algorithm in [4] uses a rational with a $j$fold pole in the shift point $\mu$,

$$r_j(\lambda) = \frac{p_j(\lambda)}{(\lambda - \mu)^j} \ .$$

Now let us study how to use a rational function with many poles,

$$r_j(\lambda) = \frac{p_j(\lambda)}{(\lambda - \mu_1)\dots(\lambda - \mu_j)} \ ,$$

determined one by one at the start of each step of the algorithm. Let us call this a Rational Krylov algorithm [19].

Our hope is that all information gathered in previous steps will help us to determine a shift in step $j$ that magnifies the interesting components and kills off those far away, even more than a shift determined in advance will do. Is that enough to offset the cost of doing several factorizations (1.2), one for each new shift $\mu_j$, instead of just one before the start?

In order to answer that question, we first develop the mechanics of the Rational Krylov recursion in section 2. We show that it gives an orthonormal basis $V$ in which the large pencil $(A, B)$ (1.1) is represented by a small Hessenberg pencil, $(K, H)$ (3.1). In section 3 we describe how to modify the basis $V$ into another orthogonal basis $W$ (3.2), from which we can compute Ritz approximations to the eigensolution as well as a suggestion for the next shift $\mu_{j+1}$. Several heuristics leading up to a usable algorithm are discussed in section 4, and we conclude by describing tests on two different applied problems in section 5.

Let me draw your attention on two observations before starting the exposition proper:

First note that we can represent a rational function with $j$ zeros and $j$ poles also as a sum of fractions,

$$r(\lambda) = \frac{p_j(\lambda)}{(\lambda - \mu_1)\dots(\lambda - \mu_j)} = \gamma_0 + \frac{\gamma_1}{\lambda - \mu_1} + \dots + \frac{\gamma_j}{\lambda - \mu_j} \ .$$

In our matrix case this is the same as starting all iterations with the same starting vector. With all shifts, $\mu_i$, $i = 1, \dots, j$, determined upon start, we can do all these matrix vector operations in parallel on $j$ processors, getting a $j$th degree polynomial in one time step.

Second, a well established technique for experimental verification of the stability of aircraft structures [21] can be described mathematically as performing several Rational Krylov iterations with different shifts but on the same starting vector. In that experiment, the aircraft prototype is suspended in a test bed and excited with an external force vector $f$ and brought to oscillate at an angular frequency $\omega$. The response vector $z$ is read off. If only structural damping is assumed we have

$$z = (-\omega^2 M + K)^{-1} f ,$$

where $M$ is a symmetric positive definite but unknown mass matrix, and $K$ is a likewise unknown but complex symmetric stiffness matrix. We see that $z$ is the result of doing one inverse iteration on the vector $f$ with shift $\omega$. This is repeated for several $\omega$ values, and a small linear combination of the obtained $z$ vectors is sought. The only differences from our technique are that the starting vector $f$ is not a part of the basis, and that measurement errors play an important role, see [22].

A word about notation: We let $A_j$ stand for a matrix with $j$ columns, the first $j$ columns of $A$ if nothing else is stated, $A_{jk}$ is a $j \times k$ matrix, and $A^{(j)}$ is the matrix as it stands after $j$ steps of the algorithm has been performed in the case some values in it gets changed later.

**2. The basic recursion.** The basic Rational Krylov recursion applied to the generalized eigenvalue problem (1.1) builds up an orthonormal basis $V$ one column at a time. In each step it first applies a shifted and inverted operator, and then makes a Gram Schmidt orthogonalization, as described in the following:

ALGORITHM RKS: *Choose starting vector* $v_1$ *and shift* $\mu_1$
*For* $j = 1, 2, \ldots$ *until* convergence

1. *Choose starting combination* $V_j t_j$
2. *Operate* $r := (A - \mu_j B)^{-1} B V_j t_j$
3. *Orthogonalize* $r := r - V_j h_j$ *where* $h_j = V_j^H r$
4. *Get new vector* $v_{j+1} := r / h_{j+1,j}$, *where* $h_{j+1,j} = \|r\|$
5. *Compute approximative solution, next shift* $\mu_{j+1}$, *and test for convergence*

We note that each step is the same as shifted and inverted Arnoldi, the only difference is that the shift $\mu$ is not the same in all steps.

Let us now see how the pencil (1.1) is expressed in the basis of $V_j$. Eliminate the intermediate vector $r$, and get

$$(A - \mu_j B)^{-1} B V_j t_j = V_{j+1} h_j ,$$

now with $h_j$ a vector of length $j + 1$ in the obvious way. Multiply by $(A - \mu_j B)$ and get

$$B V_j t_j = (A - \mu_j B) V_{j+1} h_j$$

$$AV_{j+1}h_j = BV_{j+1}(h_j\mu_j + \begin{bmatrix} t_j \\ 0 \end{bmatrix}) .$$

This is the column added at step $j$. Joining the earlier columns we get,

$$AV_{j+1}H_{j+1,j} = BV_{j+1}K_{j+1,j} , \tag{2.1}$$

where $H_{j+1,j}$ is a Hessenberg matrix built up by the orthogonalization coefficients and

$$K_{j+1,j} = H_{j+1,j}M_j + T_{j+1,j} \tag{2.2}$$

is also Hessenberg with

$$M_j = \mathrm{diag}(\mu_i) ,$$

and the triangular matrix $T_{j+1,j}$ is built up by the starting combinations.

A column normalization may be added so that

$$\begin{bmatrix} K \\ H \end{bmatrix}$$

has columns of unit length.

**3. Getting approximative eigensolution.** The basic recursion (2.1) of the preceding section computes a basis $V$ in which the pencil (1.1) is represented by the rectangular Hessenberg pencil

$$(K_{j+1,j} - \theta H_{j+1,j})s = 0 . \tag{3.1}$$

In order to get a Ritz (Galerkin) approximation to the eigensolution, we make the right matrix $H$ triangular by multiplying from the left with a unitary matrix $Q_{j+1}^H$.It consists of a product of $j$ elementary rotations zeroing the elements $(2,1), (3,2), \ldots, (j+1, j)$ one by one. Insert this into (2.1), and get

$$AV_{j+1}Q_{j+1}Q_{j+1}^H H_{j+1,j} = BV_{j+1}Q_{j+1}Q_{j+1}^H K_{j+1,j} ,$$

or

$$AW_{j+1}^{(j)}G_{j+1,j} = BW_{j+1}^{(j)}F_{j+1,j}^{(j)} , \tag{3.2}$$

now with

$$\begin{array}{rcll} W_{j+1}^{(j)} & = & V_{j+1}Q_{j+1} & \text{a new orthonormal basis} \\ G_{j+1,j} & = & Q_{j+1}^H H_{j+1,j} & \text{triangular with zero last row} \\ F_{j+1,j}^{(j)} & = & Q_{j+1}^H K_{j+1,j} & \text{a full matrix.} \end{array}$$

We use a superscript $(j)$ to distinguish intermediate quantities in step $j$ which will get changed in later steps. In the step $j+1$ the rotation will involve the $j+1$ row of $F$ and column of $W$.

The new basis $W$ is what we need, since now we can solve the square $j$ by $j$ problem

$$(F_{j,j} - \theta G_{j,j})s = 0 \tag{3.3}$$

for the Ritz value $\theta$ and Ritz vector

$$y = W_j G_{j,j} s\ . \tag{3.4}$$

For this vector the residual is

$$\begin{aligned} Ay - \theta By &= AW_j G_{j,j} s - BW_j G_{j,j} s\theta \\ &= (AW_j G_{j,j} - BW_j F_{j,j})s \\ &= Bw^{(j)}_{j+1} F^{(j)}_{j+1,\cdot} s \end{aligned}$$

This means that $y$ is a Galerkin approximation for the eigenvalue problem $B^{-1}A$ with residual

$$r = B^{-1}Ay - y\theta \tag{3.5}$$

orthogonal to the span of $W_j$.

We note that the unitary matrix $Q_{j+1}$ is of Hessenberg form. That means that the vector $w_1$ is a linear combination of $v_1$ and $v_2$, $w_2$ of $v_1$, $v_2$, and $v_3$ and so on until $w_j$ and $w^{(j)}_{j+1}$ are both combinations of all $v_1$ until $v_{j+1}$. But then we can apply the rotations, one for each step $j$, and get the $w_j$ vectors one by one and forget about the $v_j$ after that. In step 1 of ALGORITHM RKS we then use a combination $W_j^{(j-1)} t_j$ of the modified vectors and get the relation (2.2) transformed into

$$F^{(j)}_{j+1,j} = G_{j+1,j} M_j + T^{(j)}_{j+1,j}\ .$$

In all steps but the first, we have the option of choosing the last element of the vector $t_j$ as zero; then only the first column of $F$ will be changed during later steps, and $F$ will get the arrowhead form:

$$F^{(j)}_{j+1,j} = \begin{bmatrix} x & x & x & \cdots & x & x \\ x & x & x & \cdots & x & x \\ x & 0 & x & \cdots & x & x \\ \vdots & \vdots & \vdots & & \vdots & \vdots \\ x & 0 & 0 & \cdots & 0 & x \\ y & 0 & 0 & \cdots & 0 & 0 \end{bmatrix}. \tag{3.6}$$

Only the element denoted by $y$ will be changed later, and the norm of the residual (3.5) is

$$\|r\| = |ys_1|\ , \tag{3.7}$$

the product of the *first* element of $s$ and this $y$ element. Compare to the standard Arnoldi (Lanczos) case when the *last* element of $s$ is involved.

We can understand what happens if we look at the rational functions. All the vectors involved are rational functions of $C = B^{-1}A$, applied to the starting vector $v_1$. The starting vector $v_1$ corresponds to the constant term, the next vector $v_2$ has the pole $\mu_1$, and so on until the last one $v_{j+1}$ has all the poles $\mu_1, \ldots, \mu_j$. The first modified vector $w_1$ already has $\mu_1$, and the two last vectors $w_j$ and $w_{j+1}^{(j)}$ both have all the poles $\mu_1, \ldots, \mu_j$. So have both the Ritz vector $y$ (3.4) and its residual $r$ (3.5).

Let us find these rational functions. Denote the characteristic polynomial of the small problem (3.3) by

$$\chi_j(\zeta) \equiv \prod_{i=1}^{j} (\zeta - \theta_i^{(j)})$$

and the denominator with all the shifts by

$$\psi_j(\zeta) \equiv \prod_{i=1}^{j} (\zeta - \mu_i) \, ,$$

the residual (3.5) is, apart from normalization,

$$r = \chi_j(C)\psi_j(C)^{-1}v_1 \tag{3.8}$$

and the Ritz vector to $\theta_k^{(j)}$ is

$$y_k^{(j)} = \left( \prod_{i \neq k} (C - \theta_i^{(j)} I) \right) \psi_j(C)^{-1} v_1 \tag{3.9}$$

all with $C = B^{-1}A$. In the ordinary polynomial Krylov case, the Ritz vector is a polynomial of degree $j-1$ and its residual one of degree $j$, but here both have $j$ poles. This explains why all $j+1$ basis vectors $v_1, \ldots, v_{j+1}$ are present in both the Ritz vector and the residual.

**4. Algorithmic details.** In order to get a usable algorithm to compute eigenvalues we need a heuristic to determine the shift points $\mu_j$, a way of finding the starting combinations $t_j$, and an algorithm to solve the small eigenproblems (3.3) in each step.

The eigenvalues closest to the shift converge first, so we choose the starting shift $\mu_1$ in the middle of the interesting region of the complex plane. Each time we change the shift, we need to factorize the shifted matrix

$$A - \mu_j B = LU \tag{4.1}$$

using sparse Gaussian elimination, so we keep the same shift for a few steps, and only need to do forward-backward solves in step 2 of ALGORITHM RKS.

For each step $j$ we compute approximate eigenvalues from (3.3) and use their residual norms (3.7) to determine which of them that have converged. Actually we should have looked also at left residuals, in order to be able to apply the perturbation results in [9], but postpone that to a later development.

Though the matrices $F$ and $G$ are both very close to upper triangular, we use the general QZ algorithm for the small eigenproblem (3.3). However we order both rows and columns in reverse before calling QZ, to get the shifts computed from the upper left corner instead of the lower right one, and help QZ to start by getting those eigenvalues that are to converge first in the RKS algorithm. The eigenvalues will then also come in the same order from QZ when called in successive steps of RKS.

When we have used one shift $\mu$ long enough, we can choose a new $\mu_j$ from the eigenvalue approximations $\theta$ computed in step $j-1$. If no eigenvalue has converged, we take the one with the smallest residual (3.7) in the hope that it should converge faster, if some eigenvalues have converged, we take one not yet converged Ritz value that has a small residual and is in the interesting region of the complex plane.

If the shift is chosen this way, one must not be misled from the practice with the Arnoldi algorithm and choose the starting combination $t_j = e_j$, and continue with the residual $w_{j+1}$ in the next step of ALGORITHM RKS. It is the rational function (3.8) and step 2 computes the vector

$$r = (C - \mu_{j+1}I)^{-1}\chi_j(C)\psi_j(C)^{-1}v_1$$

and for $\mu_{j+1} = \theta_k^{(j)}$, the numerator and denominator gets a common factor and, cancelling those, we get $r = y_k^{(j)}$ (3.9), which is a linear combination of the columns of $W_{j+1}^{(j)}$, giving $F$ and $G$ a common nullspace in step $j+1$.

Having learnt this, we use $t_j = e_1$ when the shift is new and $t_j = e_{j-1}$ for subsequent iterations with the same shift. Then we also get the arrowhead structure (3.6) on the reduced matrix $F$.

**5. Numerical examples.** We have tested the Rational Krylov algorithm using Matlab4 [1] on SUN4 and HP 700 series workstations [2] . The linear system computations in step 2 of ALGORITHM RKS were done with the sparse matrix option in Matlab4. Reorthogonalization was done in step 3 whenever necessary. We took advantage of the complex arithmetic in Matlab even when we had real matrices. It is straightforward, but not entirely simple to make a program that preserves reality in the way indicated in [16], but we have not yet taken the effort to do it. Problems of

[1] Beta test version kindly set to our disposal by Cleve Moler, Mathworks.

[2] We are grateful to the Civil Engineering school at Chalmers for that.

sizes $n$ up to 2000 and runs up to $j = 40$ have been handled. Let us report two examples.

The first example comes from a hydrodynamical bifurcation computation that has been treated extensively in the literature, see [12], [11], and [8]. It is a parametrized nonlinear equation

$$f(x, \lambda) = 0 \tag{5.1}$$

where $x$ and $f$ are vectors of $n$ dimensions and $\lambda$ is a real parameter. We follow paths of solutions and want to localise points where the derivative matrix $f_x$ is singular, these are turning or bifurcation points.

We regard $f_x$ as a function of $\lambda$, denote it $A(\lambda)$ and assume that it is given in two points $\lambda_0$ and $\lambda_1$. Then we can predict the singular point by linear interpolation

$$A(\lambda_0 + \rho h) = A(\lambda_0) + \rho(A(\lambda_1) - A(\lambda_0)) + 1/2h^2\rho(1-\rho)R(\lambda) ,$$

where the $R$ in the rest term is bounded, and

$$h = \lambda_1 - \lambda_0 .$$

We get a linear eigenvalue problem (1.1) with $A = A(\lambda_0)$ and $B = A(\lambda_1) - A(\lambda_0)$, for which the smallest real eigenvalues $\rho$ give predictions of the positions $\lambda = \lambda_0 + \rho h$ of the singular points that are in the neighbourhood of $\lambda_0$ and $\lambda_1$. The eigenvectors give indications on how to compute bifurcating branches, see [8].

We show results on the path from $E$ to $F$ in figure 5.4 of [8]. The discretisation gives real nonsymmetric matrices of order $n = 403$. They are sparse but not very sparse, $A$ has 9883 nonzeros and its $LU$ factorization (1.2) has 29510, when we use the minimum degree ordering. The $B$ matrix is singular with rank 375, and there are many very large eigenvalues, but we are interested in the smallest real ones. The plot in figure 5.6 shows the general eigenvalue distribution when the largest ones have been peeled away.

Let us follow a typical run of ALGORITHM RKS in some detail, see figures 5.2 to 5.5! We show the a birds view of the complex plane with spikes at the Ritz approximations one unit long for each decimal judged correct. We start with a shift at zero, and keep that until the first eigenvalue converges after 10 steps. After 15 steps 3 real eigenvalues have converged, and the shift is moved down towards imaginary values. We see that the eigenvalues close to the new shift point start to grow more accurate, and after 20 steps also some of those along the negative imaginary axis have converged.

The situation after 20 steps is summarised in figure 5.7, where approximative and exact eigenvalues are plotted together. Note that the nonconverged approximations cover a region inside which essentially all eigenvalues

TABLE 5.1
*Summary of timings for numerical examples*

| Matrix | Algorithm | Conv. #eigv. | fact # | Timings | | | | Conditions | |
|---|---|---|---|---|---|---|---|---|---|
| | | | | fact | sol | rest | total | $F$ | $G$ |
| Bifurcation | RKS | 3 | 5 | 38.0 | 2.2 | 5.4 | 45.6 | 1.5e+13 | 1.0e+13 |
| | Arnoldi | 3 | 1 | 5.9 | 1.9 | 3.9 | 11.6 | 2.0e+03 | 1.1e+01 |
| Tolosa | RKS | 6 | 10 | 3.4 | 1.9 | 20.8 | 26.0 | 7.6e+16 | 1.1e+16 |
| $\mu_1 = 0$ | Arnoldi | 2 | 1 | 0.2 | 1.3 | 13.4 | 14.9 | 7.3e+03 | 1.9e+00 |
| Tolosa | RKS | 7 | 10 | 3.6 | 1.9 | 20.9 | 26.4 | 4.3e+13 | 2.9e+13 |
| $\mu_1 = 10i$ | Arnoldi | 3 | 1 | 0.4 | 1.9 | 21.0 | 23.3 | 9.6e+03 | 1.5e+02 |

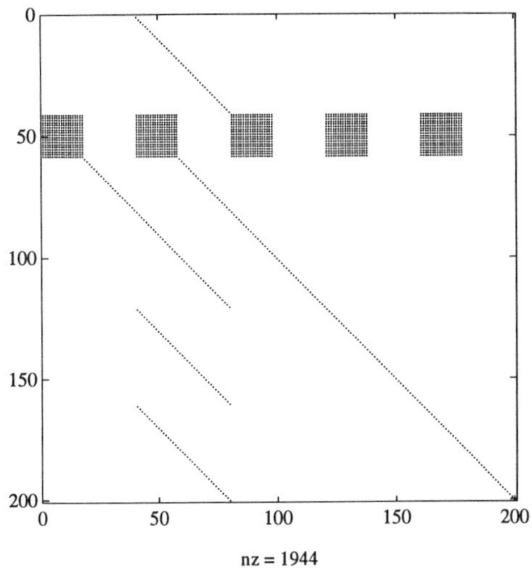


FIG. 5.1. *Profile of Tolosa matrix for* $n = 200$.

have been found. This behavior is typical, the algorithm sweeps over a region in the complex plane and finds all those eigenvalues inside. Compare this to what happens in the symmetric case [4], where the approximations for the shifted and inverted problem converge from the inside, making the approximations to the original problem come in towards the shift from the outside, all along the real line.

In order to compare our algorithm to Arnoldi with shift and invert, we ran our program with a fixed shift $\mu = 0$, and plotted the convergence patterns for the eigenvalues in figures 5.8 and 5.9. It is evident how the new shifts in RKS make some eigenvalues converge fast, while Arnoldi with fixed shift gives a more stable but slower convergence. Timings from a run on a HP 700 are summarised in table 5.1, and we see that this matrix is so full that the refactorizations are performed quite too many times in our setup.

The second test example comes from a stability computation for an

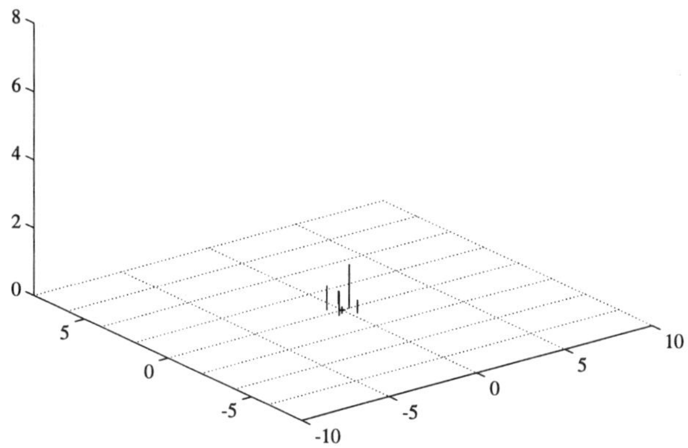


FIG. 5.2. *Bifurcation problem, approximations at j=5. Height indicates accuracy. + shift point.*

aircraft [2], named "Tolosa" after the location of the company that provided it. It looks like a Frobenius matrix and is very sparse, see the profile given in figure 5.1, where the case $n = 200$ is displayed. For larger $n$ the diagonals are just longer. We computed the spectrum for $n = 1000$ and display it in figure 5.11.

The interest is focussed on the eigenvalues close to the origin, and there is a large cluster around $\lambda = -12$, while there are a string of simple eigenvalues along the imaginary axis. It is very difficult to compute any of these eigenvalues with a direct iteration, but for an algorithm like ours that allows inversion, the only difficulty is the clustering.

We report one run with $\mu_1 = 0$, which gets some approximations in the cluster and some along the imaginary axis, and one with $\mu_1 = 10i$, which computes eigenvalues along the positive imaginary axis one by one, see figure 5.10 where we have used 30 steps to grind out 6 of the interesting eigenvalues. Timings are given in table 5.1, bottom. Here RKS is superior to Arnoldi, the refactorizations are cheap, and we get 7 eigenvalues instead of 3 in 30 steps of the algorithm.

**Acknowledgements.** Gene Golub was the one who made me aware of that my earlier version of this [19] really could be made into something, and then provided a good environment for doing that at the Linear Algebra year at the IMA in Minnesota, Spring 1992.

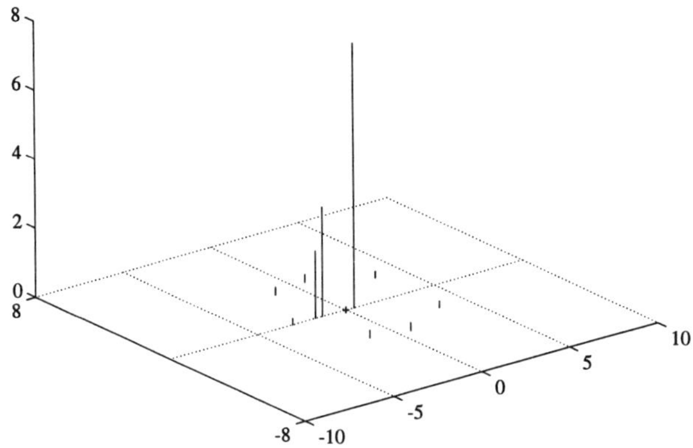


FIG. 5.3. *Bifurcation problem, approximations at j=10. One eigenvalue converged.*

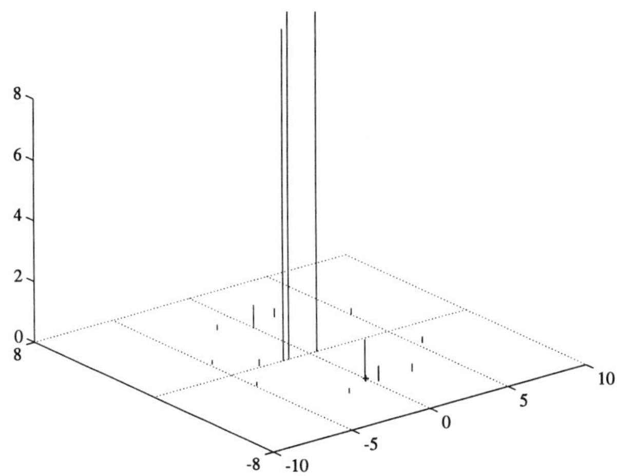


FIG. 5.4. *Bifurcation problem, approximations at j=15. 3 converged eigenvalues. Shift $+$ moved to $-4i$.*

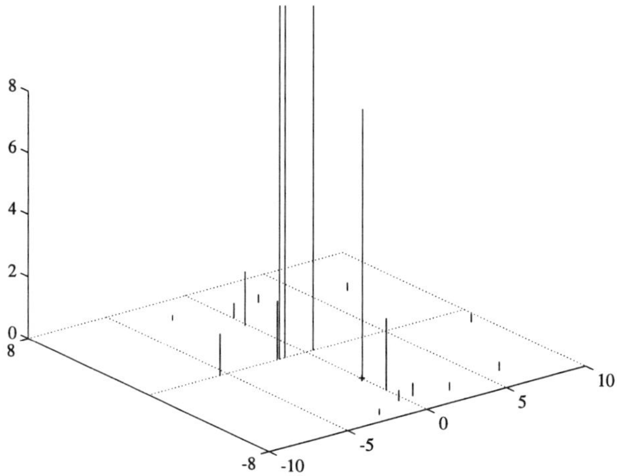


FIG. 5.5. *Bifurcation problem, approximations at j=20. 4 converged eigenvalues.*

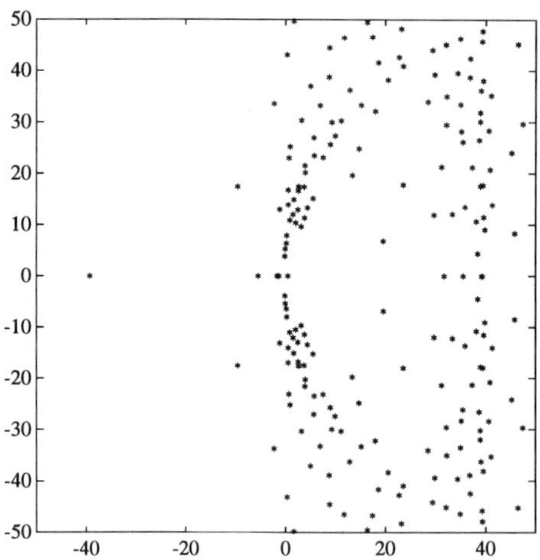


FIG. 5.6. *Bifurcation problem, overview of eigenvalues.*

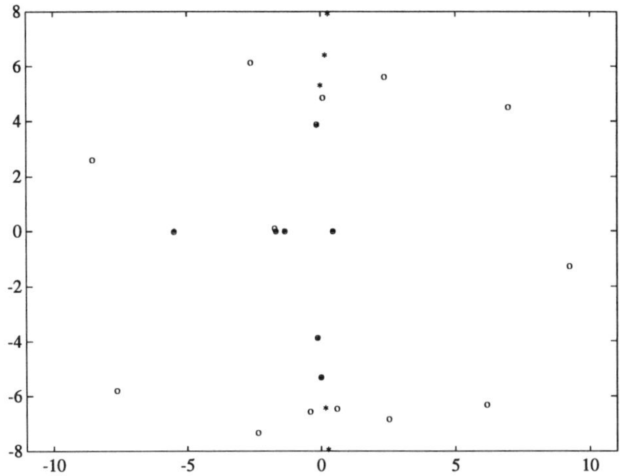


FIG. 5.7. *Bifurcation problem,* $\circ$ *computed appr at* $j = 20$, $\star$ *exact.*

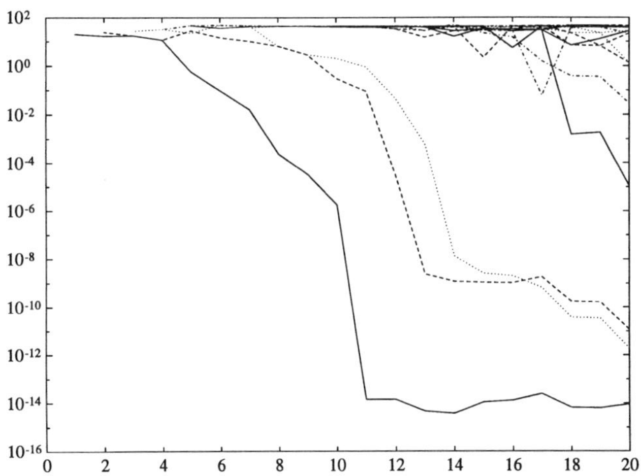


FIG. 5.8. *Bifurcation problem, convergence of eigenvalues RKS.*

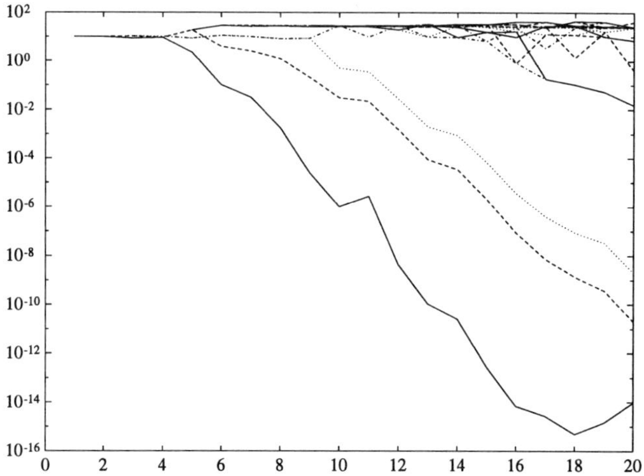


FIG. 5.9. *Bifurcation problem, convergence of eigenvalues Arnoldi.*

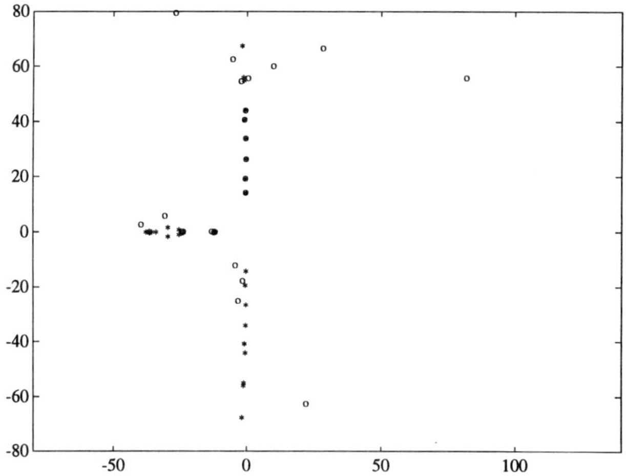


FIG. 5.10. *Tolosa matrix, Start at* $\mu_1 = 10i$, ∘ *approx,* $*$ *exact eigenv.*

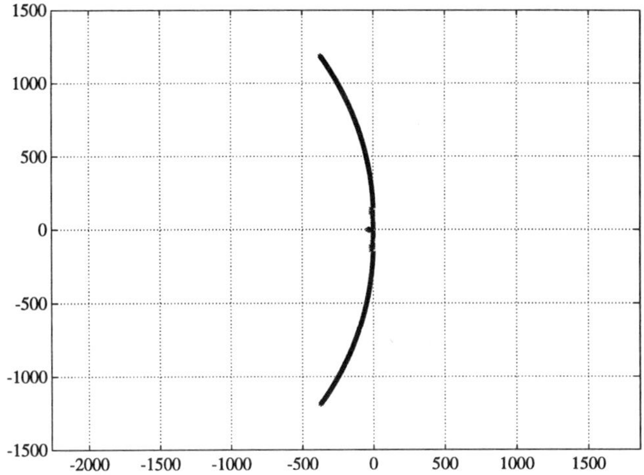


FIG. 5.11. *Tolosa matrix, all eigenvalues.*

## REFERENCES

[1] W. ARNOLDI, *The principle of minimized iterations in the solution of the matrix eigenvalue problem*, Quart. Appl. Math., 9 (1951), pp. 17–29.

[2] F. CHATELIN AND S. GODET-THOBIE. *Stability analysis in aeronautical industries, in Proceedings of the 2nd Symposium on High-Performance Computing*, Montpellier, France, M. Durand and F. El Dabaghi eds, Elsevier/North-Holland, pp.415-422, 1991.

[3] D. L. BOLEY AND G. H. GOLUB, *The nonsymmetric Lanczos algorithm and controllability*, Systems Control Lett., 16 (1991), pp. 97–105.

[4] T. ERICSSON AND A. RUHE, *The spectral transformation Lanczos method for the numerical solution of large sparse generalized symmetric eigenvalue problems*, Math. Comp., 35 (1980), pp. 1251–1268.

[5] G. GOLUB, R. UNDERWOOD, AND J. WILKINSON, *The Lanczos algorithm for the symmetric $Ax = \lambda Bx$ problem*, Tech. Report STAN-CS-72-270, Computer Science, Stanford University, Stanford, CA, 1972.

[6] R. G. GRIMES, J. G. LEWIS, AND H. D. SIMON, *A shifted block Lanczos algorithm for solving sparse symmetric generalized eigenproblems*, tech. report, Appl. Math. Stat. RTS div. Boeing Comp. Serv., M/S 7L-21, Seattle WA 98124, USA, 1991.

[7] M. H. GUTKNECHT, *A completed theory for the unsymmetric Lanczos process and related algorithms, part I*, SIMAX, 13 (1992), pp. 594–639.

[8] J. HUITFELDT AND A. RUHE, *A new algorithm for numerical path following applied to an example from hydrodynamical flow*, SISSC, 11 (1990), pp. 1181–1192.

[9] W. KAHAN, B. PARLETT, AND E. JIANG, *Residual bounds on approximate eigensystems of nonnormal matrices*, SINUM, 19 (1982), pp. 470–484.

[10] C. LANCZOS, *An iteration method for the solution of the eigenvalue problem of linear differential and integral operators*, National Bureau of Standards, Journal of Research, 45 (1950), pp. 255–282.

[11] R. MEYER-SPASCHE, *Some bifurcation diagrams for Taylor vortex flows*, Phys. Fluids, 28 (1985), pp. 1248–1252.

[12] R. Meyer-Spasche and H. B. Keller, *Computations of the axisymmetric flow between rotating cylinders*, Journ. Comp. Physics, 35 (1980), pp. 100–109.

[13] C. Paige, *Practical use of the symmetric Lanczos process with reorthogonalization*, BIT, 10 (1970), pp. 183–195.

[14] ———, *The Computation of Eigenvalues and Eigenvectors of Very Large Sparse Matrices*, PhD thesis, London University, London, England, 1971.

[15] B. Parlett, *Reduction to tridiagonal form and minimal realizations*, SIMAX, 13 (1992), pp. 567–593.

[16] B. Parlett and Y. Saad, *Complex shift and invert strategies for real matrices*, Lin. Alg. Appl., 88/89 (1987), pp. 575–595.

[17] B. Parlett and D. Scott, *The Lanczos algorithm with selective orthogonalization*, Math. Comp., 33 (1979), pp. 217–238.

[18] A. Ruhe, *The two-sided Arnoldi algorithm for nonsymmetric eigenvalue problems*, in Matrix Pencils, LNM 973, B. Kågström and A. Ruhe, eds., Springer-Verlag, Berlin Heidelberg New York, 1983, pp. 104–120.

[19] ———, *Rational Krylov sequence methods for eigenvalue computation*, LAA, 58 (1984), pp. 391–405.

[20] Y. Saad, *Variations of Arnoldi's method for computing eigenelements of large unsymmetric matrices*, Lin. Alg. Appl., 34 (1980), pp. 269–295.

[21] H. Wittmeyer, *Parameteridentifikation bei Strukturen mit benachbarten Eigenfrequenzen speziell bei Flugschwingungsversuchen*, Z. Flugwissenschaft und Weltraumforschung, 6 (1982), pp. 80–90.

[22] L. Wittmeyer-Koch, *Comparison of the evaluation of a shake test using either the $l_1$-norm or the $l_2$-norm*, Tech. Report LiTH-MAT-R-1990-36, Dept of Math. Univ. Linköping, Sweden, 1990.

# HIGHLY PARALLEL PRECONDITIONERS FOR GENERAL SPARSE MATRICES *

YOUCEF SAAD †

**Abstract.** The degree of parallelism in the preconditioned Krylov subspace method using standard preconditioners is limited and can lead to poor performance on massively parallel computers. In this paper we examine this problem and consider a number of alternatives based both on multi-coloring ideas and polynomial preconditioning. The emphasis is on methods that deal specifically with general unstructured sparse matrices such as those arising from finite element methods on unstructured grids. It is argued that multi-coloring can be combined with multiple-step relaxation preconditioners to achieve a good level of parallelism while keeping the rates of convergence to good levels. We also exploit the idea of multi-coloring and independent set orderings to introduce a multi-elimination incomplete LU factorization named ILUM, which is related to multifrontal elimination. The main goal of the paper is to discuss some of the prevailing ideas and to compare them on a few test problems.

**Key words.** Large linear systems; Krylov subspace methods; iterative methods; preconditioned conjugate gradient; Multi-coloring; incomplete *LU* preconditioning.

**1. Introduction.** Direct methods for solving linear systems are often preferred to iterative methods in real applications. For two-dimensional problems, direct solvers can be quite effective and, perhaps more importantly, they are very robust. On the other hand, the memory and computational requirements for 3-dimensional problems or 2-dimensional PDE's with many degrees of freedom per grid point, may cause serious challenges to the most efficient direct solvers available today. In addition, iterative methods are currently gaining ground because they are much easier than direct methods to implement efficiently on high-performance computers. Typical sparse direct solvers have taken decades to reach the stage of their current level of efficiency [21]. Given the complexity of these computer codes, it will take a substantial time before we reach a comparable level of efficiency on new architectures. In contrast porting iterative methods on massively parallel computers, will only require to develop a good implementation of the low level primitives, such as matrix-vector multiplications, inner products, and vector combinations, and to select a preconditioner that achieves a good compromise between the degree of parallelism and the convergence rate.

Preconditioners that allow a large degree of parallelism such as, in the simplest case, diagonal scaling, may necessitate a much larger number of iterations to converge when compared with their sequential counterparts, e.g., the standard Incomplete LU factorization (ILU). For example, a rather

* This research was supported in part by the Army Research Office under contract DAAL03-89-C-0038 and in part by a DARPA grant.

† University of Minnesota, Computer Science Department, 4-192 EE/CSci Building, 200 Union Street S.E., Minneapolis, MN 55455.

popular technique is to reorder the equations by coloring the unknowns in such a way that no two unknowns of the same color are related by an equation. In the simplest case of the 5-point matrix arising from the centered difference discretization of the Laplacean in 2 or 3 dimensional spaces, only two colors are needed, commonly referred to as red and black. If the unknowns of the same color are numbered consecutively, then a large degree of parallelism is available in the preconditioning phase. The drawback of this approach is that if an ILU type preconditioned Krylov subspace method is used then the number of iterations may increase substantially, often defeating the benefits gained from the higher degree of parallelism. Although it was observed [12] that there are orderings that have precisely the opposite effect these are not too well understood, and it is unlikely that any specific rules can be derived for the general case where the matrix arises from the discretization of coupled Partial Differential Equations on an unstructured grid as is typically the case in Computational Fluid Dynamics (CFD) for example.

Our experiments show that much can be done in recovering the good convergence properties by increasing the accuracy of the highly parallel preconditioner e.g., by performing several steps in SOR/SSOR or by using higher level of fill-in in the ILU factorization. For example, a $k$-step SSOR preconditioner applied to the Red-Black ordered matrix can achieve an excellent parallel performance. This leads us to explore the possibilities of exploiting general purpose multi-coloring techniques for general sparse matrices.

We will start by giving an overview of the standard techniques used in implementing preconditioned Krylov subspace methods. Then we will discuss some newer ideas and issues related to multi-coloring.

In order to illustrate the techniques, we found it preferable to present our numerical experiments along the way rather than at the end of the paper. Throughout the paper we will use the following three model problems.

*Problem 1.* We consider the Partial Differential Equation:

$$-\Delta u + \gamma \left( \frac{\partial e^{xy} u}{\partial x} + \frac{\partial e^{-xy} u}{\partial y} \right) + \alpha u = f ,$$

with Dirichlet Boundary Conditions and $\gamma = 10, \alpha = -60$; discretized with centered differences on a $27 \times 27 \times 27$ grid. This leads to a linear system with $N = 25^3 = 15,625$ unknowns.

*Problem 2.* We consider the same problem as above but with the parameters $\gamma = 60, \alpha = -10$.

*Problem 3.* We consider a linear system made up from matrix Sherman-3 of the Harwell-Boeing collection. This matrix is of dimension $N = 5,005$ and has $nz = 20,033$ nonzero elements. It arises in a IMPES (IMplicit Pressure, Explicit Saturation) simulation of a black-oil reservoir on a $35 \times 11 \times 13$ grid.

For all the above matrices, we construct the right hand side artificially to be of the form $b = Ae$, where the solution $e$ is the vector whose components are all ones. The initial guess is always a random vector. Although the problems are associated with regular 3-dimensional grids, this was not exploited, i.e., the matrices are considered as general unstructured sparse. These three problems are by no means easy to solve by iterative methods but they are not representative of the most difficult ones that arise in some industrial applications. We also point out that all the methods tested start with the same given data and that no additional information on the problem is exploited by any of the methods. For example, the polynomial preconditioning techniques are not supplied with spectral information or information concerning the location of the eigenvalues. Concerning the stopping criterion, all iterative solution methods are stopped as soon as the residual norm is reduced by a factor of $\epsilon = 10^{-7}$. A double star in the tables means that the method under consideration has failed to converge within the maximum number of iterations allowed. All experiments have been performed on a Cray-2 in single precision (64bits arithmetic).

**2. Preconditioned Krylov subspace methods.** We consider the linear system

$$Ax = b, \tag{2.1}$$

where $A$ is a large sparse nonsymmetric real matrix of size $N$. Several conjugate gradient-type iterative techniques based on projection processes on so-called Krylov subspaces have been proposed in recent years to solve such systems. A small list of references in this area is [8,9,13,19,18,24,22,28,29,31, 50,36,41,42] GMRES is a technique introduced in [42] for solving general large sparse nonsymmetric linear systems of equations by minimizing the 2-norm of the residual vector $b - Ax$ over $x$ in the Krylov subspace

$$K_m = \text{Span}\{r_0, Ar_0, \ldots, A^{m-1}r_0\},$$

where $r_0$ is the initial residual vector $b - Ax_0$.

The idea of preconditioning is simply to transform the above system, e.g., by multiplying it through with a certain matrix $M^{-1}$, into one that will be easier to solve by a Krylov subspace method. For example, when the preconditioner $M$ is applied to the right, we will be solving instead of (2.1), the preconditioned linear system

$$(AM^{-1})(Mx) = b. \tag{2.2}$$

In some applications it can be important to allow the preconditioner to fluctuate from step to step in the inner Krylov subspace iteration method. For later reference we now describe a variant of the GMRES algorithm based on this approach which was developed in [39]. The last step in the GMRES algorithm for solving (2.2) forms the solution as a linear combination of the preconditioned vectors $z_i = M^{-1}v_i, i = 1, \ldots, m$, where $v_i$ are the Arnoldi vectors and $M$ is the preconditioning. Because these vectors are all obtained by applying the same preconditioning matrix $M^{-1}$ to the $v$'s, we need not save them. We only need to apply $M^{-1}$ to the linear combination of the $v's$, i.e., to $V_m y_m$. If we allowed the preconditioner to vary at every step, i.e., if $z_j$ is now defined by

$$z_j = M_j^{-1} v_j$$

we may think of saving these vectors to use them in up-dating $x_m$. This observation yields the following 'flexible' preconditioning GMRES algorithm.

ALGORITHM 2.1. **FGMRES: Flexible variant of preconditioned GMRES**

1. **Start:** *Choose $x_0$ and a dimension $m$ of the Krylov subspaces. Define an $(m+1) \times m$ matrix $\bar{H}_m$ and initialize all its entries $h_{i,j}$ to zero.*
2. **Arnoldi process:**
   1. *Compute $r_0 = b - Ax_0$, $\beta = \|r_0\|_2$ and $v_1 = r_0/\beta$.*
   2. *For $j = 1, ..., m$ do*
      - *Compute $z_j := M_j^{-1} v_j$*
      - *Compute $w := Az_j$*
      - *For $i = 1, \ldots, j$, do* $\begin{cases} h_{i,j} := (w, v_i) \\ w := w - h_{i,j} v_i \end{cases}$
      - *Compute $h_{j+1,j} = \|w\|_2$ and $v_{j+1} = w/h_{j+1,j}$.*
   3. *Define $Z_m := [z_1, ...., z_m]$.*
3. **Form the approximate solution:** *Compute $x_m = x_0 + Z_m y_m$ where*
   $y_m = \text{argmin}_y \|\beta e_1 - \bar{H}_m y\|_2$ *and* $e_1 = [1, 0, \ldots, 0]^T$.
4. **Restart:** *If satisfied stop, else set $x_0 \leftarrow x_m$ and goto 2.*

For additional details, see reference [39].

**2.1. Standard preconditioners.** Apart from the trivial preconditioners such as Jacobi preconditioning (i.e., diagonal scaling), the most popular sequential preconditioners for general sparse linear systems are variants of the following three approaches.

1. Incomplete LU factorization without fill-in (ILU(0)).
2. Increased fill-in Incomplete LU (ILU) factorizations. Two distinct approaches have been used in the past. The first, level-of-fill approach, is based on the pattern only [30,51]. The second is based on numerical values and threshold strategies [40,57].
3. Relaxation type preconditioners (SOR, SSOR).

We now briefly discuss these three approaches in turn.

*ILU(0).* The Incomplete LU factorization without fill-in [30] is one of the simplest and most popular techniques in sequential machines. The principle is to find a pair of matrices $L$ and $U$, where $L$ is unit lower triangular and $U$ upper triangular such that $L+U$ has the same pattern as

$A$ and $(LU)_{ij} = a_{ij}$ for each pair $(i, j)$ that belong to the nonzero structure of $A$, i.e., for every $i, j$ so that $a_{i,j} \neq 0$. In fact such a factorization is not uniquely defined. The usual way of defining it properly is to give a constructive definition. For example: *ILU(0) is obtained by performing the standard Gaussian Elimination and replacing by zero any fill-in element during the process.* Clearly, the zeros that are introduced need not be stored.

*ILUT(p, τ).* This incomplete factorization technique is based on a two-parameter strategy for dropping elements [40]. The algorithm is based on the usual $(i, k, j)$ (row-oriented) version of Gaussian elimination, i.,e., the $i$ rows of $L$ and $U$ are determined simultaneously at step $i$. At each step the algorithm will perform the linear combination of the current version of the row (called *row* in the algorithm) with the $j$th row of $U$ if $row(k)$ is nonzero. During this combination small elements, namely those smaller that the tolerance times the norm of the original norm are dropped. ILUT$(p, \tau)$ is simply a sparse implementation of the following algorithm.

ALGORITHM 2.2. **ILUT**$(p, \tau)$

```
do i=2, n
   row(1:n) = a(i,1:n)
   nrm := || row(1:n) ||
   do  for (k=1,i-1 and where row(k) is nonzero)
       row(k) := row(k) / a(k,k)
       if (row(k) .lt. tau * nrm) then row(k) := 0
        if (|row(k)| * || u(k,k+1:n)|| .gt.  tau*nrm) then
           row(k+1:n)=row(k+1:n)-row(k)*u(k,k+1:n)
        endif
        do j=k+1,n
           if (abs(row(j)) .lt. tau*nrm) row(j) := 0
        enddo
   enddo
   l(i,1:i-1) = { p largest elements in row(1:i-1) }
   u(i,i:n) = diagonal element + {p largest elements
                                           in row(i+1:n)}
enddo
```

The scalar $\tau$ can be viewed as a parameter enabling a filtering mechanism to reduce the overall execution time and the experiments do confirm that it is a very important factor in reducing computational costs. Note that a slight variant of the above algorithm is to allow $p$ elements on the $i$-th row of $L$ *in addition* to the number of nonzero elements originally in the $i$-th row of $L$. Similarly for $U$.

*SOR/ SSOR (k).* A class of preconditioners that will be emphasized in this paper is one based on relaxation techniques. The best known of these is the SSOR preconditioning matrix defined by

$$M_{SSOR}(A) = (D - \omega E)D^{-1}(D - \omega F)$$

in which $-E$ is the strict lower part of $A$, $-F$ its strict upper part, and $D$ its diagonal.

In the context of preconditioning, it is common to simply take $\omega = 1$ as the gains from selecting an optimal $\omega$ are typically small. Note also that an optimal $\omega$ for the SOR iteration is unlikely to be the optimal $\omega$ for the combination SOR/GMRES. However, it is clear that one can use different values of $\omega$ at each step of FGMRES and this can open up the possibility of using heuristics to determine the best $\omega$ dynamically, by simply monitoring convergence.

More generally, for any iteration of the form

$$u_{k+1} = M^{-1}(Nu_k + b) \ , \quad A = M - N \tag{2.3}$$

we can define a preconditioning matrix associated with applying $k+1$ steps of this iterative process by noticing that

$$u_{k+1} = u_0 + \left[ \sum_{j=0}^{k} (M^{-1}N)^j \right] M^{-1}[b - Au_0] \tag{2.4}$$

which means that $k+1$ steps of (2.3) amount to approximating the exact solution $x = u_0 + A^{-1}[b - Au_0]$ by (2.4). Therefore the corresponding preconditioning matrix is

$$M_k = \left[ \sum_{j=0}^{k} (M^{-1}N)^j \right] M^{-1} \tag{2.5}$$

We also note that $M_k$ is nothing but the truncated Neuman series applied to

$$[M(I - M^{-1}N)]^{-1} = (I - M^{-1}N)]^{-1} M^{-1}$$

The preconditioning matrix (2.5) will be referred to as the k-step preconditioner associated with the splitting $A = M - N$. Adams [4,1] studied such preconditioners in the context of the conjugate gradient method and showed in particular that for the SSOR splitting, the preconditioning matrix is positive definite under certain conditions. Clearly, the same techniques can be applied to nonsymmetric matrices as well, although the theory is not as well understood. An important observation here is that in the non-Hermitian case, there is little attraction in using SSOR as opposed to the simpler SOR. In fact we found that for our model problems SOR(k) was generally superior to SSOR(k). We point out that the preconditioned matrix which is given by

$$M_k A = \left[ \sum_{j=0}^{k} (M^{-1}N)^j \right] M^{-1}(M - N) = I - (M^{-1}N)^{k+1} \tag{2.6}$$

will typically have eigenvalues with much better separation for larger values of $k$ than for smaller $k$, leading to fewer outer iterations to converge. However, the cost of each inner iteration is higher.

**2.2. Standard parallel implementations.** Looking at a typical conjugate gradient type algorithm we observe that the main operations are the following.

1. Setting up the preconditioner;
2. Matrix vector multiplications;
3. BLAS1 type operations: vector updates and dot products;
4. Preconditioning operations.

In the above list potential difficulties may arise in setting-up the preconditioner (1) and in the solution of linear systems with $M$, i.e., operation (4). The rest causes no major difficulties. Thus, matrix-by-vector product operations are relatively easy to implement efficiently on most computers. In the simplest case where the matrix is regularly structured, i.e., when it consists of a few diagonals this operation can be performed by multiplying diagonals with the vector [27]. The matrix can be stored in a rectangular array together with the offsets of these diagonals from the main diagonal. A number of generalizations of this formats for general sparse matrices have been proposed, the first of which is the ELLPACK-ITPACK format [32,56]. Assuming that the maximum number of nonzero elements per row $jmax$ is small we can store the entries of the matrix in a real array $C(1:n, 1:jmax)$, the $i$-th row of which contains the nonzero elements of the $i$-th row of $A$. We also need an integer array $JC(1:n, 1:jmax)$ to store the column numbers of each entry of $C$. A more general scheme is the so-called Jagged-diagonal storage format. We start by reordering the rows of the matrix in decreasing number of nonzero elements. Then, a new data structure is built by constructing what we call "jagged diagonals". The leftmost element from each row are stored along with their column indices. This is followed by the second jagged diagonal consisting of the second nonzero element in each row, and so on. The number of these jagged diagonals is equal to the number of nonzero elements of the first row, i.e., to the largest number of nonzero elements per row. These two popular schemes lead to reasonably good performance for matrix-vector products on vector machines.

An operation that may be more troublesome in preconditioned Krylov subspace methods, is the preconditioning operations. At each step, we must solve a system of the form

$$Mz = y.$$

For SSOR and for ILU type preconditioners, $M$ is the product of a sparse unit lower triangular matrix $L$ and a sparse upper triangular matrix $U$. Let us assume that the linear system (2.1) arises from a centered finite difference discretization of an elliptic partial differential equation on a rectangle using

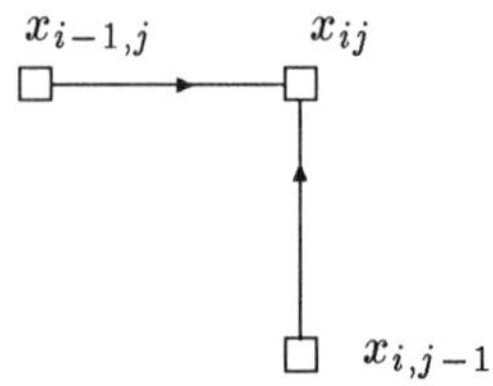


FIG. 2.1. *Stencil of the lower triangular matrix.*

an $n \times m$ mesh. This yields a linear system of size $N = nm$. For ILU(0) and SSOR the corresponding incomplete factors $L$ and $U$ have the same structure as the lower part and the upper part of the matrix $A$ respectively. If the unknowns are labeled using the natural ordering, then the 'stencil' represented in Figure 2.1 establishes the data dependency between the unknowns in the lower triangular system solution when considered from the point of view of a grid of unknowns. Specifically, in order to compute the unknown in position $(i, j)$ we only need to know the two unknowns in positions $(i-1, j)$ and $(i, j-1)$. As a result we can start computing $x_{11}$ which does not depend on any other variable, and then use this to get $x_{1,2}$ and $x_{2,1}$ simultaneously. Then these two values will in turn enable us to obtain $x_{3,1}, x_{2,2}$ and $x_{1,3}$ simultaneously and so on. The computation can be performed in wavefronts.

An important observation to make here is that the maximum degree of parallelism reached, or vector length in the case of vector processing, is the minimum of $n_x$, $n_y$ for 2-D problems. For 3-D problems the parallelism is of the order of the maximum size of the sets of domain points $x_{i,j,k}$, where $i + j + k = lev$, a constant level $lev$. See [47] for details on vector implementations. However, as can be easily seen there is little parallelism or vectorization at the beginning and at the end of the sweep. Initially the degree of parallelism is very small, starting at one then increasing to its maximum to decrease back to 1 again at the end of the sweep. These first and last few steps may take a heavy toll on achievable speed-ups on massively parallel computers.

The simple scheme described above can be generalized to irregular grids. The technique, referred to as *level scheduling* is described for example in [6,5,43,52] but we omit the details.

We now report some numerical results for our three test problems using the ILUT preconditioner with different values of the parameters $\tau$ and fill-in $p$. For Problem 1 a comparison is given with the standard sequential technique in Table 2.1. The first line shows the performs with the standard ILU(0). We note that ILU(0) is slightly different from ILUT(0, 0.0) which attempts to keep the largest elements during the incomplete factorization. However, as is shown in the optimized part of the table, the number of iterations is almost identical with that achieved with ILUT(0, 0.0001). The

TABLE 2.1
*Performance Comparison of GMRES(10)-ILUT(p,$\tau$), for problem 1, without and with the use of level-scheduling.*

| $p$ | $\tau$ | ILUT time | GMRES time | tot. time | Iter. |
|---|---|---|---|---|---|
| Problem 1, Natural ordering – Unoptimized Version | | | | | |
| ILU(0) | NA | 0.340E+00 | 0.125E+02 | 0.129E+02 | 81 |
| 1 | 0.1E-02 | 0.239E+01 | 0.160E+02 | 0.183E+02 | 90 |
| 3 | 0.1E-02 | 0.330E+01 | 0.106E+02 | 0.139E+02 | 48 |
| 5 | 0.1E-03 | 0.812E+01 | 0.787E+01 | 0.160E+02 | 30 |
| 10 | 0.1E-03 | 0.122E+02 | 0.933E+01 | 0.215E+02 | 25 |
| Problem 1, Natural ordering – Optimized Version | | | | | |
| 0 | 0.1E-03 | 0.532E+01 | 0.558E+01 | 0.109E+02 | 80 |
| 1 | 0.1E-02 | 0.241E+01 | 0.627E+01 | 0.868E+01 | 90 |
| 3 | 0.1E-02 | 0.332E+01 | 0.369E+01 | 0.701E+01 | 48 |
| 5 | 0.1E-03 | 0.808E+01 | 0.255E+01 | 0.106E+02 | 30 |
| 10 | 0.1E-03 | 0.121E+02 | 0.270E+01 | 0.148E+02 | 25 |

TABLE 2.2
*Performance of Comparison of GMRES(10)-ILUT(p,$\tau$), for problem 2, using various values of p and $\tau$, with level-scheduling.*

| Problem 2, Natural ordering – Optimized Version | | | | | | |
|---|---|---|---|---|---|---|
| $p$ | $\tau$ | ILUT time | GMRES time | tot. time | Mem. | Iter. |
| 1 | 0.100E-02 | 0.254E+01 | 0.950E+00 | 0.349E+01 | 151016 | 13 |
| 3 | 0.100E-02 | 0.302E+01 | 0.903E+00 | 0.392E+01 | 207418 | 11 |
| 5 | 0.100E-02 | 0.337E+01 | 0.862E+00 | 0.423E+01 | 241435 | 10 |
| 7 | 0.100E-02 | 0.364E+01 | 0.909E+00 | 0.455E+01 | 264305 | 10 |
| 5 | 0.100E-03 | 0.117E+01 | 0.123E+01 | 0.240E+01 | 51465 | 43 |
| 10 | 0.100E-03 | 0.169E+01 | 0.761E+00 | 0.245E+01 | 75822 | 20 |
| 15 | 0.100E-03 | 0.212E+01 | 0.935E+00 | 0.305E+01 | 96745 | 19 |

TABLE 2.3
*Performance of Comparison of GMRES(10)-ILUT(p,$\tau$), for problem 3, using various values of p and $\tau$, with level-scheduling.*

| Problem 3 – Optimized Version of ILUT | | | | | | |
|---|---|---|---|---|---|---|
| $p$ | $\tau$ | ILUT time | GMRES time | tot. time | Mem. | Iter. |
| 1 | 0.100E-02 | 0.531E+00 | 0.350E+01 | 0.403E+01 | 30060 | 158 |
| 2 | 0.100E-02 | 0.578E+00 | 0.320E+01 | 0.378E+01 | 35217 | 139 |
| 3 | 0.100E-02 | 0.630E+00 | 0.255E+01 | 0.318E+01 | 40174 | 105 |
| 4 | 0.100E-02 | 0.679E+00 | 0.235E+01 | 0.303E+01 | 44864 | 92 |
| 5 | 0.100E-02 | 0.723E+00 | 0.184E+01 | 0.256E+01 | 49240 | 68 |
| 7 | 0.100E-02 | 0.768E+00 | 0.205E+01 | 0.282E+01 | 56965 | 69 |
| 10 | 0.100E-02 | 0.819E+00 | 0.204E+01 | 0.286E+01 | 63324 | 61 |
| 5 | 0.100E-03 | 0.117E+01 | 0.123E+01 | 0.240E+01 | 51465 | 43 |
| 10 | 0.100E-03 | 0.169E+01 | 0.761E+00 | 0.245E+01 | 75822 | 20 |
| 15 | 0.100E-03 | 0.212E+01 | 0.935E+00 | 0.305E+01 | 96745 | 19 |

iteration part of an optimized ILU(0)-GMRES(10) should therefore take about the same time as that with ILUT(0,0.0001) - GMRES(10) shown on the second part of the table. We also point out that the meaning of $p$ is slightly different for that in [40]. We allow $p$ fill-ins *in addition* to the number of nonzero elements in the lower part of $A$. Similarly with the upper part. In all the tests below the dimension of the Krylov subspace in GMRES is taken to be equal to 10. As is to be expected the number of iterations required for convergence decreases as the accuracy of the preconditioner increases. However, the cost of setting up the ILUT factorization increases rapidly as does the memory cost. In the 'optimized' part of the tables, level scheduling is used to optimize the forward and backward solves. In addition, the matrix vector multiplications that arise in the algorithm as well as in the level-scheduled forward and backward solutions are optimized by using what is referred to as the "jagged diagonal" format [38,6]. The numbers in the 'memory' columns reported for Problems 2 and 3, represent the memory locations needed to store the ILUT factorization only. No attempt has been made to optimize the preprocessing phase and as a result the ILUT computation is likely to run close to scalar speed. Some optimization can be achieved but the returns are likely to be limited, at least with this approach. It is interesting to note that even if we disregard the ILUT time, the speed-up achieved here is modest. A factor of at most three is gained in the iteration part of the solution.

**3. Polynomial preconditioning.** When vector computers first became available, polynomial preconditioners were among the first alternatives proposed to the standard ILU preconditioning techniques [7,25,35,26,44,53]. These methods consist of choosing a polynomial $s$ and replacing the original linear system by

$$s(A)Ax = s(A)b \quad \text{or} \quad A(s(A)y) = b, \quad x = s(A)y \ , \tag{3.1}$$

which is then solved by a conjugate gradient type technique. The motivation here is that the only operations that are needed with the matrix are matrix by vector products.

The determination of an optimum polynomial $s$ is guided by a requirement to have the preconditioned matrix $As(A)$ as close as possible to the identity matrix in some sense. This is achieved by exploiting the fact that we may adaptively construct a complex domain that roughly represents the spectrum of the matrix $A$. If we assume that the spectrum $\sigma(A)$ is contained in some continuous set $E$ that encloses it, then a good choice for the polynomial $s$ is obtained by solving the problem,

$$\begin{aligned} &\text{Find } s \in \Pi_k \text{ which minimizes:} \\ &\max_{\lambda \in E} |1 - \lambda s(\lambda)|. \end{aligned} \tag{3.2}$$

Several possible choices for $E$ have been used in the literature. The simplest idea is to use an ellipse $E$ [28,29] that encloses an approximate convex hull of the spectrum. Then the shifted and scaled Chebyshev polynomials are nearly optimal, and even optimal in some special instances [17]. A second alternative is to use a polygon $H$ that contains $\sigma(A)$ [46,36]. The motivation here is that polygons may better represent the shape of an arbitrary spectrum. The polynomial is not explicitly known but it may be computed iteratively by a Remez-type algorithm.

An alternative considered by Johnson et al. in [25] for the Hermitian case and generalized to non-Hermitian matrices in [36] is to use a least squares polynomial instead of the infinity norm polynomial. We would then need to solve

$$\text{Find } s \in \Pi_k \text{ that minimizes:} \qquad \qquad (3.3)\quad \|1 - \lambda s(\lambda)\|_w ,$$

where $w$ is some weight function on the boundary of $H$ and $\|.\|_w$ is the $L_2$-norm associated with the corresponding $L_2$ inner product. In [36] we used an $L_2$-norm associated with Chebyshev weights on the edges of a polygon $H$ containing the spectrum and expressed the best polynomial as a linear combination of Chebyshev polynomials associated with the ellipse of smallest area containing $H$. If the contour of $H$ consists of $\mu$ edges each with center $c_i$ and half-length $d_i$, then the weight on each edge is defined by

$$w_i(\lambda) = \frac{2}{\pi}\,|d_i - (\lambda - c_i)^2|^{-1/2}. \qquad (3.4)$$

With these weights, or any other Jacobi weights on the edges, there is a finite procedure to compute the best polynomial *that does not require numerical integration*; for details see [36]. An advantage of the least squares approach over the Chebyshev approach is its better robustness properties. In addition, there are common simple regions in the complex plane which ellipse cannot represent well. Thus, it is also more general. Its disadvantages are that it involves more arithmetic and that the maximum degree that can be used is limited, mostly due to stability constraints.

In the symmetric positive definite case, where $H$ is reduced to an interval, it is not too difficult to obtain an approximate interval $H$ containing the spectrum by exploiting the relationship between the Lanczos algorithm and the conjugate gradient method; see e.g., [11,22]. In the non-Hermitian case a number of techniques can be used similarly. We next describe an implementation based on a combination with GMRES [42]. Eigenvalue estimates can be computed from the Hessenberg matrices generated from GMRES. Thus, the idea is to use a certain number, say $m_1$, of steps of GMRES to get eigenvalue estimates and to improve the current solution. The eigenvalue estimates are used to improve the current polygon $H$ and to compute the next least squares polynomial. A number of steps (say $m_2$) of GMRES are then performed for the preconditioned system $s(A)Ax = s(A)b$. This procedure is outlined below.

ALGORITHM 3.1. **Polynomial Preconditioned GMRES**

1. **(Re-) Start:** *Compute current residual vector* $r := b - Ax$.
2. **Adaptive GMRES run:** *Run* $m_1$ *steps of GMRES for solving* $Ad = r$. *Update* $x$ *by* $x := x + d$. *Get eigenvalue estimates from the eigenvalues of the Hessenberg matrix.*
3. **Compute new polynomial**. *Refine* $H$ *by using new eigenvalue estimates. Get new best polynomial* $s_k$.
4. **Polynomial Iteration:**
   - *Compute the current residual vector* $r = b - Ax$.
   - *Run* $m_2$ *steps of GMRES applied to* $As_k(A)q = r$. *Update* $x$ *by* $x := x + s_k(A)q$.
   - *Test for convergence. If solution converged then stop else goto 1.*

Note that several enhancements to the above procedure can be made. First, the degree of the polynomial need not be fixed. We can start with a small degree polynomial and increase the degree as the convex hull becomes more and more accurate. In addition, we can combine polynomial preconditioning with some other preconditioning technique, such as one of the ones to be described later which offer a large degree of parallelism. 224z In the current version of our experimental Polynomial preconditioned GMRES (PGMR) code, we have not implemented the variable degree and use only diagonal preconditioning on the matrix $A$. In addition, our code sets $m_2 = m_1$.

Since the matrix-by-vector operation is the only sparse operation in PGMR, it is important to optimize it in order to achieve good overall performance. We used the jagged diagonal storage scheme [38,37] for this purpose.

In the tables, $m$ is the dimension of the Krylov subspaces, i.e., $m = m_1 = m_2$. The variable *deg* refers to the degree $k$ of the polynomial used in the above algorithm, while *its* is the number of outer iterations, and *matvecs* is the total number of matrix by vector products needed to achieve convergence. One observation we can make here is that the times obtained for different values of the parameters $m$ and *deg* show substantial variations. It is our experience that this behavior is quite common with polynomial preconditionings. Indeed, a poor estimation of the convex hull, may lead the iteration to proceed for many steps before it can correct $H$ and reach a good convergence rate. On the other hand if several systems must be solved with the same matrix, then the information about the convex hull can be exploited for subsequent systems. Observe also that the best times are comparable to those obtained with the optimized ILUT preconditioners. We note that for problem 3, the performance was rather poor for small values of $m$ and we omit the corresponding results. In general the larger $m$

the better the approximation of the convex hull $H$ obtained. However, the solution of the larger eigenvalue problems start to contribute significantly to the total times for larger $m$'s.

TABLE 3.1
*Performance of GMRES(m) with least-squares polynomial preconditioning for Problem 1.*

| Polynomial preconditioning – Problem 1 | | | | |
|---|---|---|---|---|
| $m$ | deg. | total CPU | matvecs | its |
| 10 | 5 | 0.227E+01 | 323 | 6 |
| 10 | 10 | 0.760E+01 | 1190 | 10 |
| 10 | 15 | 0.547E+01 | 870 | 5 |
| 15 | 5 | 0.141E+01 | 183 | 2 |
| 15 | 8 | 0.104E+01 | 147 | 2 |
| 15 | 10 | 0.112E+01 | 164 | 1 |
| 15 | 20 | 0.133E+01 | 214 | 1 |
| 20 | 5 | 0.148E+01 | 178 | 2 |
| 20 | 10 | 0.118E+01 | 169 | 1 |
| 20 | 15 | 0.146E+01 | 214 | 1 |
| 20 | 20 | 0.157E+01 | 239 | 1 |

TABLE 3.2
*Performance of GMRES(m) with least-squares polynomial preconditioning for Problem 2.*

| Polynomial preconditioning – Problem 2 | | | | |
|---|---|---|---|---|
| m | deg. | total CPU | matvec's | its |
| 10 | 5 | 0.914E+00 | 132 | 3 |
| 10 | 10 | 0.109E+01 | 168 | 2 |
| 10 | 15 | 0.152E+01 | 243 | 2 |
| 15 | 5 | 0.763E+00 | 102 | 2 |
| 15 | 10 | 0.839E+00 | 124 | 1 |
| 15 | 15 | 0.877E+00 | 134 | 1 |
| 20 | 5 | 0.848E+00 | 104 | 1 |
| 20 | 10 | 0.860E+00 | 119 | 1 |
| 20 | 15 | 0.976E+00 | 139 | 1 |

TABLE 3.3
*Performance of GMRES(m) with least-squares polynomial preconditioning for Problem 3.*

| Polynomial preconditioning – Problem 3 | | | | |
|---|---|---|---|---|
| m | deg. | total CPU | matvec's | its |
| 15 | 20 | 0.230E+01 | 1590 | 5 |
| 20 | 5 | 0.181E+01 | 883 | 8 |
| 20 | 10 | 0.223E+01 | 1324 | 6 |
| 20 | 15 | 0.178E+01 | 1126 | 4 |
| 20 | 20 | 0.175E+01 | 1157 | 3 |

**4. The Red-Black ordering.** A commonly used approach for solving large sparse linear systems on parallel computers is to resort to the red-black ordering which is the simplest form of the general multi-color orderings to be described later. In its simplest form the general problem that

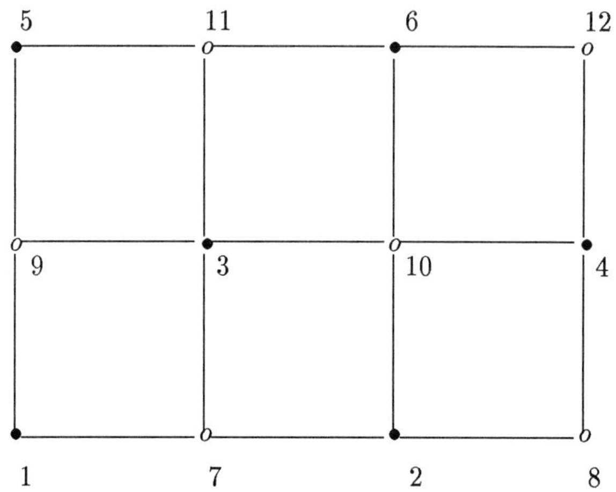


FIG. 4.1. *Red black labeling of a* $4 \times 3$ *grid.* $\bullet$ = *black node,* $o$ = *red node.*

is addressed by multi-coloring techniques is to color the nodes of a graph so that neighboring vertices have different colors. In the case of a simple 2-dimensional finite difference grid (5-point operator) only two colors (referred to as Red and Black) are needed. This 2-color (red-black) ordering is illustrated in Figure 4.1 for a $4 \times 3$ grid.

If we reorder the unknowns in such a way as to list the red unknowns first together and then the black ones, we will obtain a system of the form

$$\begin{pmatrix} D_1 & E \\ F & D_2 \end{pmatrix} \begin{pmatrix} x_1 \\ x_2 \end{pmatrix} = \begin{pmatrix} b_1 \\ b_2 \end{pmatrix} \tag{4.1}$$

where $D_1$ and $D_2$ are diagonal matrices. Matrices that can be permuted into the above form are said to have property $A$ [55]. Once the system has been reduced to the convenient form (4.1) a number of different techniques can be used for solving such systems. We now examine some of these in turn and discuss their advantages and disadvantages.

**4.1. ILU preconditionings on the reduced system.** Probably the simplest method that can be used for solving (4.1) is the standard ILU(0)

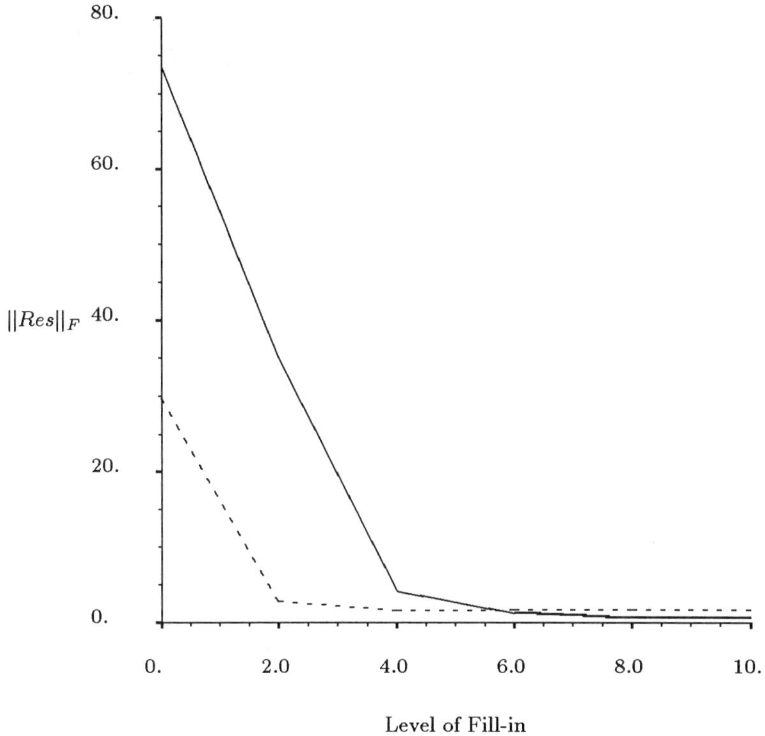


FIG. 4.2. *Frobenius norm of error matrix for the ILUT preconditioner as the accuracy increases. Dashed line: natural ordering, Solid line: Red-Black ordering.*

on this block system. The degree of parallelism here is of order $N$. A drawback is that often the number of iterations is higher than with the natural ordering, but the approach may still be competitive for *easy* problems.

Based on this, one can raise the interesting question as to whether or not the number of iterations can be reduced back to a competitive level by using a more accurate ILU factorization on the red-black system, e.g., ILUT [40]. Some recent experiments reveal that the answer is yes. In fact the situation is often reversed in that for the same level of fill-in $p$, the red-black ordering will outperform the natural ordering preconditioner for $p$ large enough, in terms of number of iterations. A look at the plots of the Frobenius norms for the residual matrices $A - LU$, for a test matrix similar to that of model problems 1 and 2, reveals that indeed the incomplete factorization may become more accurate for the red-black matrices, as $p$ increases, see Figure 4.2. The dashed line shows the Frobenius norms of the errors $||A - LU||_F$ for the natural ordering and the solid line shows the same norms for the Red-Black ordering using the same fill-in level. Notice that when the level of fill-in reaches 6, the quality of the RB preconditioner as measured by the F-norm of the error becomes better than that of the natural ordering using the same level of fill $p$. This reverses the situations obtained for smaller values of the fill-in level $p$.

Unfortunately, one notable difficulty with this approach, is that fill-ins will be introduced in the (2,2) blocks in $L$ and $U$ which loose their diagonal structure. As a result the degree of parallelism is severely reduced.

Table 4.1 shows the performance of GMRES(10)-ILUT(p,$\tau$), for various values of $p$ and $\tau$ for solving Problem 1 after the red-black ordering is applied. Similarly to Table 2.1, level scheduling is used to optimize the forward and backward solves and all the matrix vector multiplications are performed using the "jagged diagonal" format [38,6]. Again, the preprocessing needed to compute the incomplete factorization itself is not optimized.

TABLE 4.1
*Performance comparison of GMRES(10)-ILUT(p,$\tau$), for various values of p and $\tau$, without and with the use of level-scheduling. Red-Black ordering.*

| $p$ | $\tau$ | ILUT time | GMRES time | tot. time | Iter. |
|---|---|---|---|---|---|
| Red-Black ordering – Unoptimized Version | | | | | |
| 1 | 0.1E-02 | 0.182E+01 | 0.258E+02 | 0.276E+02 | ** |
| 3 | 0.1E-02 | 0.248E+01 | 0.248E+02 | 0.273E+02 | 136 |
| 5 | 0.1E-03 | 0.414E+01 | 0.184E+02 | 0.225E+02 | 90 |
| 10 | 0.1E-03 | 0.704E+01 | 0.724E+01 | 0.143E+02 | 28 |
| Red-Black ordering – Optimized Version | | | | | |
| 1 | 0.1E-02 | 0.186E+01 | 0.108E+02 | 0.127E+02 | ** |
| 3 | 0.1E-02 | 0.254E+01 | 0.970E+01 | 0.122E+02 | 136 |
| 5 | 0.1E-03 | 0.419E+01 | 0.683E+01 | 0.110E+02 | 90 |
| 10 | 0.1E-03 | 0.712E+01 | 0.255E+01 | 0.967E+01 | 28 |

Note that ILU(0) and MILU(0) were also tried in this example but they failed to converge as did ILUT(1,0.001) (The maximum number of steps allowed in GMRES is 160 in this test). In general, the times achieved are higher than those with the natural ordering for the low accuracy preconditioners (p=1 and p=3). However the situation improves drastically for the more accurate versions of ILUT and the conclusion on the comparison with the natural ordering is reversed for $p = 10$, i.e., the red-black ordering now outperforms the natural ordering. We have made the same observation in many other examples. An additional interesting feature here is that the ILUT time is also reduced with the red-black ordering.

**4.2. Multiple step SOR/SSOR preconditioners.** As is suggested above for ILUT, high accuracy preconditioners on the Red-Black matrix can reduce the number of iterations required for convergence. However the fill-ins introduced in the "black" part of the system with such high level ILUs ruin the degree of parallelism achieved from the red-black ordering. A remedy against this shortcoming is to resort to similar *high-level SOR* *or SSOR preconditioners* instead of ILUT. This is a significant advantage of relaxation type preconditioners. We can define high level SOR/SSOR preconditioning by performing several SOR or SSOR steps in each preconditioning operation, instead of just one as is traditionally done. This can be combined with the flexible version of GMRES described earlier if needed.

In Table 4.1 we show the performance of a GMRES(m)-SOR(k) strategy based on this approach for Problem 1. The $m$ parameter takes 2 different values $m = 10$ and $m = 20$. The $k$ parameter, the number of steps in each SOR preconditioning takes several different values. Note that for $k = 1$ and $m = 10$ convergence is not achieved. An important observation we can make in this experiment and many other similar ones is that the optimal $k$ for a fixed $m$ is often achieved for values much larger than one.

TABLE 4.2
*Performance of Red-Black SOR(k)–GMRES(m), multiple-step Gauss-Seidel Preconditioned GMRES for Problem 1, for different values of m and k.*

| $m$ | $k$ | Tot time | Iter. time | Iter. |
|---|---|---|---|---|
| 10 | 1 | 0.377E+01 | 0.357E+01 | ** |
| 10 | 5 | 0.240E+01 | 0.220E+01 | 60 |
| 10 | 10 | 0.205E+01 | 0.185E+01 | 30 |
| 10 | 15 | 0.229E+01 | 0.208E+01 | 21 |
| 10 | 20 | 0.225E+01 | 0.205E+01 | 18 |
| 10 | 30 | 0.185E+01 | 0.165E+01 | 10 |
| 10 | 35 | 0.206E+01 | 0.186E+01 | 10 |
| 20 | 1 | 0.246E+01 | 0.226E+01 | 139 |
| 20 | 5 | 0.154E+01 | 0.134E+01 | 36 |
| 20 | 10 | 0.124E+01 | 0.104E+01 | 18 |
| 20 | 15 | 0.146E+01 | 0.127E+01 | 15 |
| 20 | 20 | 0.164E+01 | 0.145E+01 | 13 |

We note that with $m = 20$ and $k = 10$, this approach is about 2.5 times faster that the best possible performance that we could obtain from the standard ILU preconditioners. If we compared the total times then the new approach would be even more attractive. Another attractive feature is that the preprocessing – coloring and forming the reduced system and coloring again – is highly parallel and quite easy to optimize whereas high level-of-fill ILU-based preconditioner offer a limited degree of parallelism.

Figure 4.3 shows the iteration times and total times obtained for Problem 2 as $k$ varies from 1 to 30. Here the dimension $m$ of the Krylov subspace is fixed to 10. Note that the preprocessing time which is roughly constant and equal to 0.2 second becomes nonnegligible for this problem because the iteration times are smaller than for Problem 1. Again we should emphasize that the preprocessing is a highly parallel process but it has not been not optimized in any way for these experiments. We show in Table 4.2 the results of a few runs for Problem 1 and, similarly, in Table 4.3 some results

TABLE 4.3
*SOR(k)-GMRES(m), multiple-step Gauss-Seidel Preconditioned GMRES, for different values of m and k for problems 2 and 3.*

| $m$ | $k$ | Tot time | Iter.time | Iter. |
|---|---|---|---|---|
| Problem 2 | | | | |
| 10 | 1 | 0.157E+01 | 0.137E+01 | 100 |
| 10 | 5 | 0.691E+00 | 0.492E+00 | 13 |
| 10 | 10 | 0.657E+00 | 0.459E+00 | 7 |
| 10 | 15 | 0.720E+00 | 0.520E+00 | 5 |
| 20 | 5 | 0.703E+00 | 0.505E+00 | 12 |
| 20 | 10 | 0.671E+00 | 0.471E+00 | 7 |
| 20 | 15 | 0.723E+00 | 0.522E+00 | 5 |
| 20 | 20 | 0.866E+00 | 0.664E+00 | 4 |
| Problem 3 | | | | |
| 20 | 5 | 0.316E+01 | 0.311E+01 | 261 |
| 20 | 10 | 0.494E+01 | 0.480E+01 | 238 |
| 20 | 15 | 0.579E+01 | 0.565E+01 | 184 |

for Problems 2 and 3. In all the tables the total times include the times for permutations and conversion to the Ellpack-Itpack format required for optimizing the Red-black SOR iterations.

A second possibility not explored in this paper is to use block versions of the SOR iteration such as a line SOR relaxation.

**4.3. Reduced system approach.** A second method that has been used in conjunction with the red-black ordering is to eliminate the red unknowns by forming the reduced system which involves only the black unknowns, namely,

$$(D_2 - FD_1^{-1}E)x_2 = b_2 - FD_1^{-1}b_1.$$

This linear system is again sparse and has about half as many unknowns. It has been observed that for 'easy problems' this reduced system can often be efficiently solved with only diagonal preconditioning. The preprocessing to compute the reduced system is highly parallel and inexpensive. In addition the reduced system is usually well-conditioned and has some interesting properties when the original system is highly nonsymmetric [15]. We should point out that it is not necessary to form the reduced system. This latter strategy is more often employed when $D_1$ is not diagonal, such in Domain Decomposition methods, but it can also have some uses in other situations. For example, applying the matrix to a given vector $x$ can be achieved using nearest-neighbor communication in the original finite difference grid and can be performed efficiently without forming the Schur complement matrix. In addition this can save storage, which may be more critical in some applications.

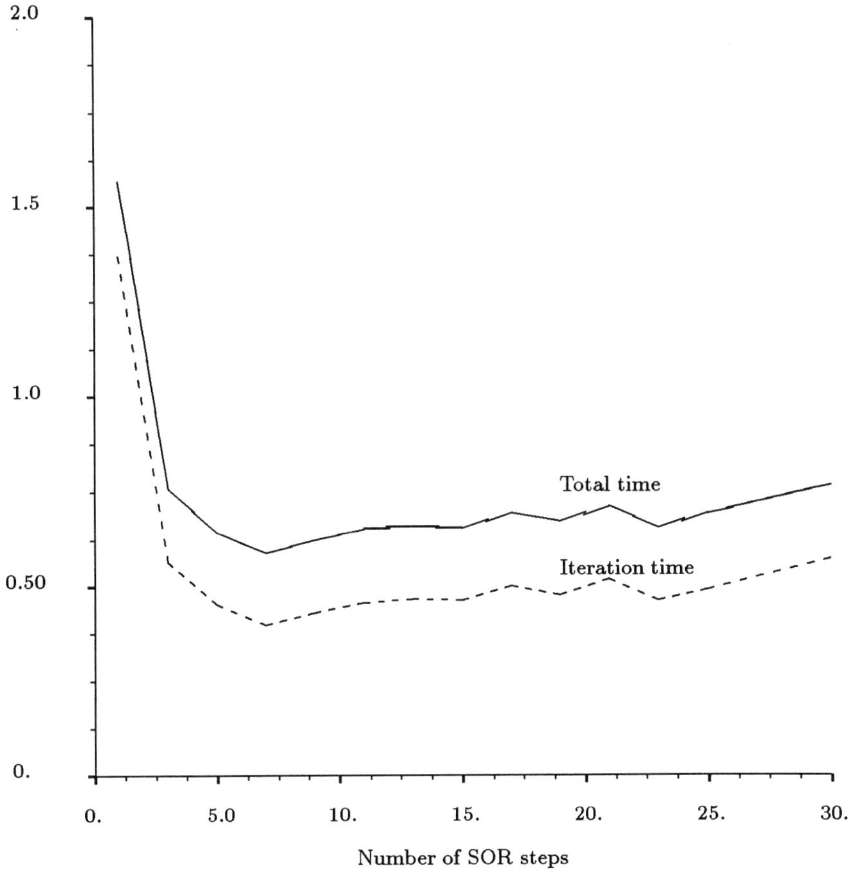


FIG. 4.3. *Iteration and total times for SOR(k)-GMRES(10) as k varies.*

We experimented with problems 1, 2, and 3, using this approach. A two-coloring of the matrix is done and then GMRES(m) is used with only diagonal preconditioning to solve the reduced system. The jagged diagonal data structure is used to get a good performance for the matrix-vector products which are the only sparse matrix operations required. The stopping criterion we used is to require the same condition on the residual norm for this reduced system as for the original problem. It is clear that the test on the reduced system does not guarantee that the corresponding solution to the original system will satisfy the same condition. Table 4.4 shows the times obtained with $m = 10$ and $m = 20$ for both problems. The iteration time is the time needed for GMRES to converge for the reduced system. The rest of the time includes the time for coloring, reducing the system, converting into jagged diagonal data structure, and back-substituting to obtain the 'red' unknowns. We observe that for Problem 3, the method essentially failed to converge and similarly for Problem 1, convergence is achieved for larger values of $m$ only. In general when convergence is achieved the CPU times are much higher than with other techniques seen earlier. As a general rule, it seems that an approach that is based on a simple diagonal preconditioned GMRES iteration applied to the reduced system is not reliable.

TABLE 4.4
*Performance comparison of a reduced system approach using diagonal preconditioning GMRES(m) solver.*

| $m$ | Total time | GMRES time | GMRES steps |
|---|---|---|---|
| Problem 1 | | | |
| 10 | 0.101E+02 | 0.837E+01 | ** |
| 20 | 0.607E+01 | 0.436E+01 | 134 |
| Problem 2 | | | |
| 10 | 0.442E+01 | 0.268E+01 | 80 |
| 20 | 0.366E+01 | 0.194E+01 | 59 |
| Problem 3 | | | |
| 10 | 0.217E+01 | 0.191E+01 | ** |
| 20 | 0.220E+01 | 0.195E+01 | ** |

**5. Multi-coloring.** Generalizations of the red-black ordering have often been used in the context of PDE's as a means of introducing parallelism. The goal of multi-coloring is to color the vertices of an arbitrary adjacency graph in such a way that two adjacent nodes do not have a common color. The optimality issue, getting large color sets or small numbers of colors, is secondary. In fact in the context of iterative methods keeping preprocessing costs to a minimum is essential and in this framework, a sub-optimal multi-coloring is generally sufficient. The general technique of multi-coloring has been used in particular for understanding the theory of relaxation tech-

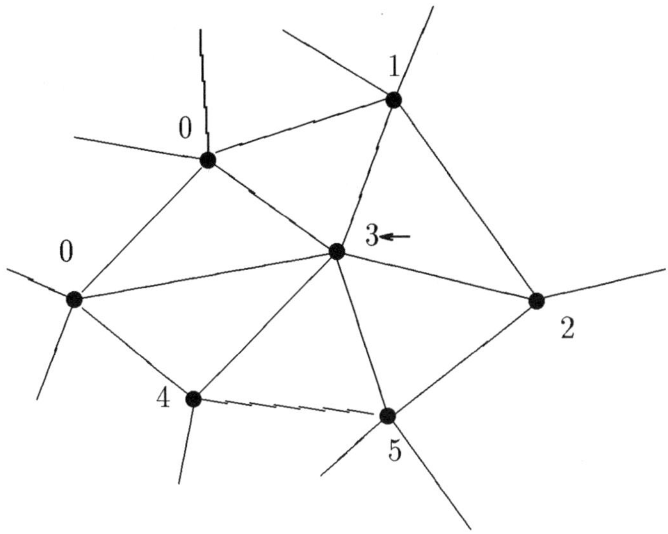


FIG. 5.1. *Illustration of the greedy approach for multi-coloring. The node being assigned a color is indicated by an arrow. It will be assigned color number 3, the smallest of the allowable color numbers, which are the positive integers other than 1, 2, 4, 5.*

niques [55,48] as well as for deriving efficient alternative formulations of some relaxation algorithms [48,20]. More recently, it was emerged as a useful tool for introducing parallelism in iterative methods, see for example [4,3,34,14,33]. Multi-Coloring is also commonly employed in a slightly different form – coloring elements (or edges) as opposed to nodes – in finite elements (or finite volume) techniques [10,49]. Multi-coloring is especially useful in element-by-element techniques when forming the residual, i.e., in multiplying an unassembled matrix by a vector. The contributions of the elements of the same color can all be evaluated and applied simultaneously to the resulting vector [23,16,45]. We start by describing a simple technique for multi-coloring a graph. Then we will consider the same questions we addressed for the red-black ordering regarding ways of solving multi-colored systems.

**5.1. A greedy multi-coloring algorithm.** A basic and quite simple algorithm for obtaining a multi-coloring of an arbitrary graph, relying on a greedy approach can be sketched as follows.

1. Initially assign color number zero (uncolored) to every node.
2. Choose an order in which to traverse the nodes.
3. Scan all nodes in the chosen order and at every node $i$ do

$$\text{Color}(i) = \min\{k > 0 \mid k \neq \text{Color}(j), \forall\, j \in \text{Adj}(i)\},$$

As before, Adj($i$) represents the set of nodes that are adjacent to node $i$. The color assigned to node $i$ in step 3 is simply the smallest allowable color number which can be assigned to node $i$, where *allowable* means different from the colors of the neighbors and positive.

A few simple facts concerning this greedy algorithm, are the following.

- The initial order of traversal may be important in reducing the number of colors.
- If a graph is two-colorable, the algorithm will find the common 2-color (Red-Black) ordering for *Breadth-First* traversals or *Depth-First* traversals.
- The number of colors needed does not exceed the maximum degree of each node +1.

The parallelization of the algorithm is important and can be achieved in one of two ways. First, we can use the same algorithm and resort to some form of locking mechanism. The node being colored scans all neighbors in turn, and locks them (possibly after they are unlocked by some other node). Then it reads their colors and assigns a color number to its node and finally unlocks its neighbors. A second strategy, is to use a divide-and-conquer, or 'domain-decomposition' approach: recursively color the subdomains – then the nodes in the interfaces.

The idea of the greedy multi-coloring algorithm is known in Finite Element techniques (to color elements), see e.g., Berger-Brouays-Syre (1982), and Benantar and Flaherty. Wu [54] presents the greedy algorithm for multi-coloring vertices and uses it for SOR type iterations. Finally, the effect of multi-coloring has been extensively studied by Adams [4,2] and Poole and Ortega [34].

**5.2. Multiple step SOR / SSOR preconditioners.** Just as for the red black ordering, we can use ILU0 or SOR, SSOR preconditioning on the reordered system. The parallelism of SOR/ SSOR is still of order $N$ in general, but again one can anticipate a loss in the efficiency as the number of iterations may increase.

One important point of detail here concerns the implementation of the multicolor SOR technique. It is clear that similarly to the Red-Black ordering a Gauss-Seidel sweep will essentially require $p$ matrix-vector products, where $p$ is the number of colors. Specifically, let us assume that the matrix is stored in the well-known ELLPACK-ITPACK format and that the block structure of the permuted matrix is defined by a pointer array *iptr*. The index $iptr(j)$ is the number of the first row in the $j$-th block. Thus, the pair $A(n1 : n2, *), JA(n1 : n2, *)$ represents the sparse matrix consisting of the rows $n1$ to $n2$ in the Ellpack format. The main diagonal of $A$ is assumed to be stored separately in inverted form in a one-dimensional array *diag*. One single step of the multicolor SOR iteration will then take the following form.

ALGORITHM 5.1. **Multi-color SOR sweep in the ELLPACK format**

```
0. Comment: does one sweep of Gauss-Seidel for solving
                                                    A y = rhs.
1.   do kol = 1, nkol               | nkol = number of colors
2.      n1 = iptr(kol)               | beginning of block # kol
3.      n2 = iptr(kol+1)-1           | end of block # kol
4.      y(n1:n2) = rhs(n1:n2)
5.      do j=1,ndiag
6.         do i = n1, n2
7.             y(i) = y(i) - a(i,j)*y(ja(i,j))
8.         enddo
9.      enddo
6.      y(n1:n2) = diag(n1:n2) * y(n1:n2)    | D**(-1) * result
7.   enddo
```

In the above code, the outer loop, i.e., the loop in line 1, is sequential. The loop starting in line 6, is vectorizable / parallelizable. Clearly more parallelism can still be extracted in the combination of the loops in lines 5 and 6. For example, on the CM-2 using the CM FORTRAN, Thinking Machines' implementation of FORTRAN-90, these two loops can be rewritten in a one-line instruction as

```
y(n1:n2)=y(n1:n2)-sum(a(n1:n2,1:ndiag)*spread(y(ja(n1:n2),1,
                                                    n2-n1+1),2)
```

In the above instruction the product of arrays is a component-wise product. The *spread* basically reproduces the vector $n2 - n1 + 1$ times in the first dimension whereas the *sum* 'reduces along the second dimension', i.e., it will sum all the components of the resulting array along the second dimension.

Going back to our three model problems, we note that we have already discussed the reduced system approach, in section 4.3. We have shown in Table 4.2 some timing results for the case where the reduced systems are solved with a simple-minded diagonal preconditioned GMRES iteration. We now show in Table 5.1 the analogue of Table 4.2 for the case where the reduced system is solved using a multicolored SOR(k) preconditioning. Thus, the reduced system obtained from the first coloring is multicolored again and reordered. In addition, the final reordered matrix obtained is processed to be stored in the Ellpack format in order to achieve a good performance in the SOR iteration as is described in Algorithm 5.1. Note that the preprocessing is higher than with the previous approach which required no reduction but only a coloring and a permutation. In the present case, we need to color, permute, reduce, color, and permute again. In addition, at the very last step, we also convert the permuted matrix into the Ellpack-Itpack format. These permutation and coloring operations have not been optimized and as they stand are highly sequential operations.

Similar results regarding Problem 2 and Problem 3, are given in Table 5.2 and Table 5.3 respectively. A general observation here is that the iteration times (GMRES times) are much lower than those seen in previous experiments.

TABLE 5.1
*Performance of multicolor SOR(k)-GMRES(m) for Problem 1, using multiple-step Gauss-Seidel preconditioning for the reduced system – one level of reduction.*

| $m$ | $k$ | Total time | Iter. time | Iter. |
|---|---|---|---|---|
| 10 | 1 | 0.306E+01 | 0.119E+01 | 69 |
| 10 | 5 | 0.291E+01 | 0.102E+01 | 20 |
| 10 | 10 | 0.284E+01 | 0.102E+01 | 11 |
| 10 | 15 | 0.308E+01 | 0.125E+01 | 9 |
| 10 | 20 | 0.321E+01 | 0.136E+01 | 8 |
| 10 | 25 | 0.334E+01 | 0.156E+01 | 7 |
| 20 | 1 | 0.270E+01 | 0.804E+00 | 41 |
| 20 | 3 | 0.257E+01 | 0.705E+00 | 20 |
| 20 | 5 | 0.268E+01 | 0.792E+00 | 15 |
| 20 | 10 | 0.288E+01 | 0.101E+01 | 11 |
| 20 | 15 | 0.308E+01 | 0.122E+01 | 9 |
| 20 | 20 | 0.338E+01 | 0.143E+01 | 8 |
| 20 | 25 | 0.355E+01 | 0.165E+01 | 7 |

TABLE 5.2
*Problem 2. One level of reduction, GMRES(m)-SOR(k) used to solve the reduced system.*

| $m$ | $k$ | Total time | Iter. time | Iter. |
|---|---|---|---|---|
| 10 | 1 | 0.220E+01 | 0.352E+00 | 20 |
| 10 | 2 | 0.204E+01 | 0.254E+00 | 10 |
| 10 | 3 | 0.209E+01 | 0.230E+00 | 7 |
| 10 | 4 | 0.211E+01 | 0.273E+00 | 6 |
| 10 | 5 | 0.211E+01 | 0.257E+00 | 5 |
| 10 | 6 | 0.213E+01 | 0.254E+00 | 4 |
| 10 | 8 | 0.203E+01 | 0.254E+00 | 3 |
| 10 | 10 | 0.213E+01 | 0.310E+00 | 3 |
| 10 | 15 | 0.220E+01 | 0.339E+00 | 2 |

**5.3. Multi-level reduced Systems.** When we form the reduced system associated with the elimination of the unknowns of the first color we obtain a system that is again sparse. An idea that comes immediately to mind is to color the resulting system again and form the reduced system associated with the elimination of the first color. We will call this second reduced system the second-level reduced system. The process can be continued recursively a few times. However, as the level of the reduction increases, the reduced systems become denser. Nevertheless this approach can be successful if a small number of reductions are used and if the last reduced system is solved efficiently, e.g., by a multi-color SOR precondi-

TABLE 5.3
*Problem 3. One level of reduction, GMRES(m)-SOR(k) used to solve the reduced system.*

| $m$ | $k$ | Total time | Iter. time | Iter. |
|---|---|---|---|---|
| 10 | 1 | 0.143E+01 | 0.114E+01 | ** |
| 10 | 5 | 0.209E+01 | 0.180E+01 | 139 |
| 10 | 10 | 0.261E+01 | 0.232E+01 | 99 |
| 10 | 15 | 0.289E+01 | 0.260E+01 | 76 |
| 20 | 1 | 0.141E+01 | 0.112E+01 | 241 |
| 20 | 5 | 0.124E+01 | 0.954E+00 | 76 |
| 20 | 10 | 0.157E+01 | 0.128E+01 | 58 |
| 20 | 15 | 0.176E+01 | 0.148E+01 | 46 |

tioned technique. In addition, a variant of it will lead to a general purpose ILU preconditioner that can be viewed as a parallel version of ILUT which is the subject of a separate section.

We will now illustrate the multi-level reduced system technique in detail. Figure 5.2 shows the stages required in a 2-level reduction. On top from left to right we show the original matrix, and the matrix obtained after the first coloring and permutation. In this case the original matrix is a $100 \times 100$ matrix obtained from a $10 \times 10$ centered difference discretization of the Laplacean. So the multi-coloring algorithm yielded the usual Red-Black coloring in this case. In the middle we show on the left the reduced matrix obtained by eliminating the first half of the unknowns of the previous matrix. Then we multicolor the reduced matrix to obtain the permuted matrix on the right. We need 4 colors in this case. Finally at the bottom, we show on the left the matrix obtained by eliminating the unknowns associated with the first of the 4 colors in the previous matrix. The resulting matrix is multi-colored again to obtain the final permuted matrix shown on the right. The number of colors needed for this last matrix is 10. There are two reductions and three colorings in this example. The last coloring is only useful in case a multicolor SOR or SSOR preconditioning scheme will be applied to solve the last reduced system, or if the process of reduction were to be continued for another level. The sizes of the two reduced matrices are 50 and 35 respectively.

In the following description the right hand side is conceptually the last column of the $n \times (n+1)$ matrix $A$. If we call $A_j$ the matrix obtained at the $j$- step of the reduction, $j = 1, \ldots, nlev$ then we first multi-color it, and then we permute the matrix according to the multicolor ordering to get a matrix in the form

$$P_j A_j P_j^T = \begin{pmatrix} D_j & F_j \\ E_j & X_j \end{pmatrix} \tag{5.1}$$

where $D_j$ is a diagonal matrix. It is clear that the numbering of the colors

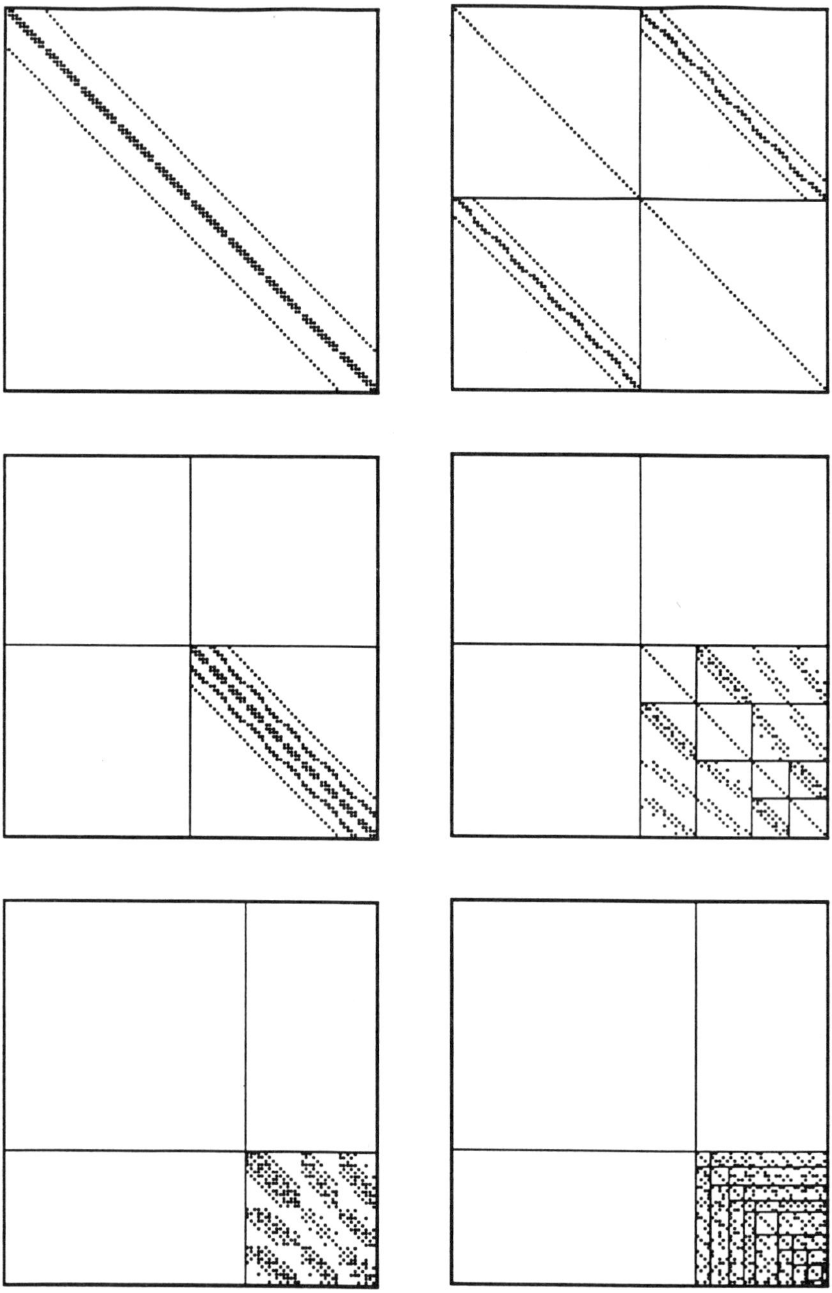

Fig. 5.2. *The MRS approach for a 5-point matrix of size* $100 \times 100$.

is arbitrary. We can select the color yielding the largest color set to be the first color. Finally, we reduce the system by eliminating the unknowns of the first color, to get the next reduced matrix via the formula,

$$A_{j+1} = X_j - E_j D_j^{-1} F_j \ . \tag{5.2}$$

The transformations used in the elimination process, i.e., the $E_j$'s, need not be saved but the upper $F_j$ matrices must be since they will be used to back-solve for the remaining equations. Regarding the matrix reordering, we can either permute the matrix or simply keep a permutation array.

ALGORITHM 5.2. **Multilevel Reduced System (MRS) Solution Algorithm**

*1. Let $A_0 = A$;*

*2. Do for $j = 0, 1, \ldots, nlev - 1$*

*Find a multi-coloring of the graph of $A_j$*

*Call color 1, the color with the largest number of elements.*

*Permute the matrix according to the multicolor ordering into the form (5.1).*

*Reduce the system by eliminating color number 1, i.e., get the system (5.2).*

*enddo*

*3. Solve the last reduced system with, e.g., GMRES with multistep-multicolor*

*SOR preconditioning.*

*4. Do for $j = nlev - 1, ..., \ldots, 0$*

*Backsolve for $A_j$ from the known solution for $A_{j+1}$.*

*enddo*

One of the advantages of this approach comes from the fact that the last matrix for which we must solve a linear system is usually much smaller than the original matrix. This, compounded with the fact that it also usually takes fewer steps for the higher level iterations to converge, makes the scheme quite attractive. The drawbacks are its complexity and the fact that the reduced matrices will eventually become quite dense. Beyond a few (say 2 or 3) levels, further reductions are no longer cost-effective. This motivates the search for an incomplete factorization technique which will basically consist of dropping small elements just as is done for the standard Gaussian Elimination. Before this, we would like to show a number of experiments to compare these techniques using our three model problems.

Some of the results obtained are summarized in Table 5.4. The numbers in the column labeled 'memory' show the memory requirement the 'factorization' alone. Comparing with the results seen in Section 4.3 which correspond to a level of reduction equal to $nlev = 1$, we can observe that in general the times are higher. In addition the memory requirement may become prohibitive for larger problems. This strongly suggests using dropping as a means of reducing the memory cost, as is done in the standard ILU techniques.

TABLE 5.4
*Two levels of reduction, GMRES(m)-SOR(k) used to solve the reduced system.*

| *nlev* | *m* | *k* | tot CPU | its CPU | memory | itstot |
|---|---|---|---|---|---|---|
| | | | Problem 1 | | | |
| 2 | 10 | 1 | 0.853E+01 | 0.464E+01 | 399129 | 72 |
| 2 | 10 | 5 | 0.750E+01 | 0.366E+01 | 399129 | 19 |
| 2 | 10 | 10 | 0.805E+01 | 0.426E+01 | 399129 | 11 |
| | | | Problem 2 | | | |
| 2 | 10 | 1 | 0.486E+01 | 0.105E+01 | 399129 | 17 |
| 2 | 10 | 5 | 0.477E+01 | 0.895E+00 | 399129 | 4 |
| 2 | 10 | 10 | 0.485E+01 | 0.970E+00 | 399129 | 2 |
| | | | Problem 3 | | | |
| 2 | 10 | 1 | 0.648E+01 | 0.315E+01 | 66107 | 250 |
| 2 | 10 | 5 | 0.890E+01 | 0.563E+01 | 66107 | 132 |
| 2 | 10 | 10 | 0.109E+02 | 0.758E+01 | 66107 | 95 |

**5.4. ILUM: a Multi-elimination ILU factorization.** As was suggested above, the successive reduced systems become more and more expensive to form and store as the number of levels increases. An alternative to cut memory and preprocessing costs is to use a simple dropping strategy as we form the reduced systems. For example, we can drop any fill-in element introduced, including pivot elements, whenever its size is less than a tolerance times the 2-norm of the original row.

Thus, a modified version of Algorithm 5.2 can be used as a preconditioner, i.e., it will provide an approximate solution $M^{-1}v$, for any given $v$. The modification will consist of the following changes. First, in the preprocessing phase, dropping is used when constructing the reduced system (5.2). This means that (5.2) is replaced by

$$A_{j+1} = X_j - E_j D_j^{-1} F_j + R_j \quad (5.3)$$

in which $R_j$ is the matrix of the elements that have been dropped. We note that in (5.3) and throughout this section it is no longer assumed that the right-hand-side is the last column of the matrix.

Second, the linear systems to be solved in the last reduced system, i.e., in step 3 of Algorithm 5.2, need no longer be solved accurately. In fact we can solve it according to the level of tolerance that we have allowed in the preprocessing phase, although we do not know how to actually choose an optimal residual norm stopping criterion. Observe that since we would like to solve the linear system inaccurately, it is imperative to use the flexible variant of GMRES described in Section 2. Alternatively, we could simply use a fixed number of multi-color SOR iterations. The implementation of the ILUM preconditioner (ILU with Multi-elimination) corresponding to this strategy is rather complicated and involves several parameters. The detailed description of the algorithm as well as some analysis and additional experiments will be the object of a forthcoming paper.

However, we would like to summarize the description of the the two phases of the algorithm, namely the preprocessing phase and the forward-backward solution phase. Essentially, the preprocessing phase consists of steps 1, and 2 of Algorithm 5.2 except that the operations related to the right-hand-side are ignored and, instead, the matrices $E_j$ obtained in (5.1) after the coloring and permutation are now also saved since they will be needed for each forward sweep in the preconditioning operations. Thus we have a succession of block-LU factorizations of the form,

$$(5.4)\quad P_j A_j P_j^T = \begin{pmatrix} D_j & F_j \\ E_j & X_j \end{pmatrix} = \begin{pmatrix} I & O \\ E_j D_j^{-1} & I \end{pmatrix} \times \begin{pmatrix} D_j & F_j \\ O & A_{j+1} \end{pmatrix}$$

with $A_{j+1}$ defined by (5.3). We denote by $x_j$ the right hand side associated with the matrix $A_j$, and recursively, $x_{j+1}$ the transformed right hand side associated with the matrix $A_{j+1}$ and call $E_j'$ the matrix $E_j' = F_j D_j^{-1}$. We can either permute the vector at each level as is suggested by the above factorizations, or we can multiply all permutations together during the factorization phase so that we will only have to apply the permutation at the beginning and the reverse permutation at the end of the backward/forward solution process. Then the forward and backward and solutions consist of an algorithm with the following structure.

ALGORITHM 5.3. *LU solve for ILUM preconditioner*

1. *Apply permutation to right-hand-side* $x$.
2. *Forward solution. For* $j = 0, 1, \ldots, nlev - 1$ *do:* $x_{j+1} := x_{j+1} - E_j' x_j$
3. *Solve last reduced system with a relative tolerance* $\epsilon$ *to get* $x_{nlev}$.
4. *Backward solution. For* $j = nlev - 1, \ldots, 0$ *do:* $x_j := D_j^{-1}(x_j - F_j x_{j+1})$.
5. *Permute the solution vector back to the original ordering.*

We now report a few experiments with a preliminary implementation of ILUM. In all the experiments the innermost reduced system is solved by SOR(1)-GMRES(20) and the outer accelerator is GMRES(10). As usual *nlev* is the number of levels in the reduction. The parameter $\tau$-ILUM shown in the second column of the tables is the dropping tolerance $\tau$ used in the preconditioning. The parameter $tol - prec$ in the third column is the tolerance used in the GMRES(20)-SOR(1) iteration of the innermost reduced linear system to solve. Again, the numbers in the column labeled 'memory' show the memory requirement the ILUM 'factorization' alone. The numbers $its - out$ refer to the number of outer iterations required, i.e., the number of ILUM preconditioned GMRES iterations to converge. The numbers $its - tot$ refer to the total number of inner iterations required, i.e., the overall total number of matrix vector products in the calls to GMRES(20)-SOR(1) needed to solve all the reduced systems occurring throughout the run to solve the linear system.

TABLE 5.5
*Performance of GMRES(10)-ILUM preconditioning for Problem 1*

| Parameters | | | Performance | | | | |
|---|---|---|---|---|---|---|---|
| *nlev* | $\tau$-ILUM | tol prec | tot CPU | its CPU | memory | its-out | its-tot |
| 1 | 0.0001 | 0.100 | 4.224 | 2.335 | 107891 | 9 | 70 |
| 2 | 0.0001 | 0.100 | 9.141 | 5.563 | 142591 | 9 | 69 |
| 3 | 0.0001 | 0.100 | 12.939 | 7.349 | 166454 | 9 | 78 |
| 1 | 0.0001 | 0.010 | 3.395 | 1.502 | 107891 | 4 | 56 |
| 2 | 0.0001 | 0.010 | 7.216 | 3.704 | 142591 | 4 | 52 |
| 3 | 0.0001 | 0.010 | 11.061 | 5.409 | 166454 | 5 | 65 |
| 1 | 0.0010 | 0.010 | 4.193 | 2.223 | 107891 | 4 | 85 |
| 2 | 0.0010 | 0.010 | 10.419 | 6.812 | 142512 | 6 | 116 |
| 3 | 0.0010 | 0.010 | 11.826 | 6.854 | 158788 | 7 | 115 |
| 1 | 0.0010 | 0.100 | 4.179 | 2.323 | 107891 | 9 | 70 |
| 2 | 0.0010 | 0.100 | 8.127 | 4.690 | 142512 | 10 | 63 |
| 3 | 0.0010 | 0.100 | 9.609 | 4.678 | 158788 | 10 | 64 |
| 1 | 0.0100 | 0.010 | 4.390 | 2.487 | 107891 | 5 | 103 |
| 2 | 0.0100 | 0.010 | 7.841 | 4.600 | 131543 | 20 | 109 |
| 3 | 0.0100 | 0.010 | 12.009 | 7.769 | 149984 | 36 | 143 |

TABLE 5.6
*Performance of GMRES(10)-ILUM preconditioning for Problem 2.*

| Parameters | | | Performance | | | | |
|---|---|---|---|---|---|---|---|
| *nlev* | $\tau$-ILUM | tol prec | tot CPU | its CPU | memory | its-out | its-tot |
| 1 | 0.0001 | 0.010 | 3.127 | 1.292 | 107745 | 5 | 41 |
| 2 | 0.0001 | 0.010 | 5.617 | 2.262 | 141167 | 4 | 30 |
| 3 | 0.0001 | 0.010 | 8.374 | 3.337 | 158741 | 4 | 30 |
| 1 | 0.0010 | 0.001 | 2.777 | 0.916 | 106592 | 3 | 33 |
| 2 | 0.0010 | 0.001 | 5.526 | 2.307 | 136035 | 4 | 40 |
| 3 | 0.0010 | 0.001 | 7.510 | 2.750 | 152345 | 4 | 37 |
| 1 | 0.0010 | 0.010 | 3.065 | 1.117 | 106592 | 4 | 33 |
| 2 | 0.0010 | 0.010 | 5.168 | 1.896 | 136035 | 4 | 29 |
| 3 | 0.0010 | 0.010 | 7.503 | 2.633 | 152345 | 5 | 33 |
| 1 | 0.0010 | 0.100 | 3.861 | 1.938 | 106592 | 10 | 35 |
| 2 | 0.0010 | 0.100 | 6.713 | 3.321 | 136035 | 10 | 35 |
| 3 | 0.0010 | 0.100 | 8.799 | 3.886 | 152345 | 11 | 32 |
| 1 | 0.0100 | 0.100 | 4.165 | 2.411 | 97870 | 12 | 55 |
| 2 | 0.0100 | 0.100 | 5.395 | 2.538 | 114047 | 12 | 37 |
| 3 | 0.0100 | 0.100 | 6.959 | 2.916 | 127401 | 13 | 40 |

TABLE 5.7
*Performance of GMRES(10)-ILUM preconditioning for Problem 3.*

| Parameters | | | Performance | | | | |
|---|---|---|---|---|---|---|---|
| *nlev* | $\tau$-ILUM | tol prec | tot CPU | its CPU | memory | its-out | its-tot |
| 1 | 0.0001 | 0.100 | 1.471 | 1.197 | 19438 | 6 | 232 |
| 2 | 0.0001 | 0.100 | 3.250 | 2.794 | 25595 | 8 | 246 |
| 3 | 0.0001 | 0.100 | 2.962 | 2.300 | 30076 | 8 | 175 |
| 4 | 0.0001 | 0.100 | 3.231 | 2.350 | 34479 | 9 | 185 |
| 5 | 0.0001 | 0.100 | 3.456 | 2.392 | 38760 | 10 | 179 |
| 10 | 0.0001 | 0.100 | 3.482 | 1.412 | 53650 | 10 | 95 |
| 15 | 0.0001 | 0.100 | 4.238 | 1.287 | 64381 | 11 | 67 |
| 20 | 0.0001 | 0.100 | 4.738 | 1.037 | 72382 | 10 | 40 |
| 25 | 0.0001 | 0.100 | 5.242 | 0.995 | 78454 | 10 | 37 |
| 1 | 0.0010 | 0.100 | 2.868 | 2.594 | 19432 | 16 | 473 |
| 2 | 0.0010 | 0.100 | 3.171 | 2.737 | 25180 | 25 | 260 |
| 3 | 0.0010 | 0.100 | 2.993 | 2.395 | 29562 | 32 | 199 |
| 4 | 0.0010 | 0.100 | 4.186 | 3.463 | 32341 | 57 | 205 |
| 5 | 0.0010 | 0.100 | 3.930 | 3.063 | 35900 | 48 | 174 |

TABLE 5.8
*Best performances obtained for each of the preconditioners tested for Problems 1 and 3. The ranking is based on the iteration time only, i.e., it ignores all preprocessing costs.*

| Method | Total time | Iteration time |
|---|---|---|
| Problem 1 | | |
| Non-optimized ILU(0)-GMRES(10) | 12.5 | 12.9 |
| ILUT-GMRES | 10.6 | 2.55 |
| Poly-GMRES | 1.04 | 1.04 |
| Red-Black SOR(k)-GMRES(m) | 1.24 | 1.04 |
| MRS – SOR(k)-GMRES(m) | 2.57 | 0.71 |
| ILUM – GMRES(m) | 3.39 | 1.50 |
| Problem 3 | | |
| Non-optimized ILU(0)-GMRES(10) | 5.79 | 7.13 |
| ILUT-GMRES | 2.40 | 1.23 |
| Poly-GMRES | 1.81 | 1.81 |
| Red-Black SOR(k)-GMRES(m) | 3.16 | 3.11 |
| MRS / SOR(k)-GMRES(m) | 1.24 | 0.954 |
| ILUM – GMRES(m) | 5.24 | 0.995 |

**6. Summary of results and conclusion.** Table 5.8 is a summary of the best performances obtained in previous tables with each of the preconditioners tested for Problems 1 and 3. It is only indicative of achievable performance with each of the preconditioners and should not in any way be taken as an absolute measure of comparison since the tables are not exhaustive. The reason why we did not include the preprocessing times in the comparisons is that they have not been optimized as yet. When preprocessing should be accounted for then polynomial preconditioning is certainly quite attractive except for the fact that performance can be variable. Here are a few observations we can make regarding these tests.

- Despite a rather preliminary implementation, ILUM is very competitive for Problem 3, the hardest to solve among the three problems. Using ILUM with 25 levels of reduction gives close to the best time achieved for this problem.
- For problems 1 and 2, the one-level reduced system (no dropping) and multi-color SOR performed best. In addition the preprocessing cost for the MRS approach with one level of reduction is modest.
- The polynomial preconditioning approach is attractive because of the negligible preprocessing cost involved. In fact the only additional cost incurred lies in the computation of the least squares polynomial.

- For the multi-level reduced system approach without dropping, it seems that the one level reduction performs usually better than with levels higher than one. It is also usually better than the zero-level (i.e. multicolor SOR preconditioned GMRES).
- One difficulty with the better performing preconditioners, is that they tend to involve more parameters. This seems to be one of the unavoidable difficulties with parallel algorithms, simply because they are often more complex than the sequential ones.
- In general the preprocessing is highly parallel in all the algorithm seen, except for the standard ILUT, so the costs reported for these times can be dramatically reduced in a parallel environment.

We would like to conclude by observing that in order to develop efficient massively parallel preconditioners it seems important to mix strategies that offer a high degree of parallelism, e.g., multi-color orderings, with strategies that increase the accuracy of the preconditioners, e.g., k-step SOR. The higher accuracy tends to compensate for the poorer convergence rates that might otherwise result from the reordering.

Our next step is to implement these ideas on a massively parallel computer such as the Thinking Machine CM-5. We expect that, generally speaking, similar conclusions will hold true. However, one potential difficulty may be the implementation of the preprocessing phase which is rather complicated for the better performing preconditioners. This may make simpler techniques such as polynomial preconditionings or the SOR(k) – SSOR(k) preconditionings combined with just one level of reduction look quite attractive overall.

**Acknowledgement.** The author would like to acknowledge the support of the Minnesota Supercomputer Institute which provided the computer facilities and an excellent research environement to conduct this research.

## REFERENCES


[1] L. Adams. M-step preconditioned conjugate gradient methods. *SIAM J. Sci. Statist. Comp*, 6:452–463, 1985.

[2] L. Adams and H. Jordan. Is SOR color-blind? *SIAM J. Sci. Statist. Comp*, 6:490–506, 1985.

[3] L. Adams and J. Ortega. A multi-color SOR Method for Parallel Computers. In *Proceedings 1982 Int. Conf. Par. Proc.*, pages 53–56, 1982.

[4] L. M. Adams. *Iterative algorithms for large sparse linear systems on parallel computers*. PhD thesis, Applied Mathematics, University of Virginia, Charlottsville, VA, 22904, 1982. Also NASA Contractor Report 166027.

[5] E. C. Anderson. Parallel implementation of preconditioned conjugate gradient methods for solving sparse systems of linear equations. Technical Report 805, CSRD, University of Illinois, Urbana, IL, 1988. MS Thesis.

[6] E. C. Anderson and Y. Saad. Solving sparse triangular systems on parallel computers. *International Journal of High Speed Computing*, 1:73–96, 1989.

[7] S. F. Ashby. *Polynomial Preconditioning for Conjugate Gradient Methods*. PhD thesis, Computer Science Dept. , University of Illinois, Urbana, IL, 1987. Available as Technical Report 1355.

[8] S. F. Ashby, T. A. Manteuffel, and P. E. Saylor. Adaptive polynomial preconditioning for Hermitian indefinite linear systems. *BIT*, 29:583–609, 1989.

[9] O. Axelsson. A generalized conjugate gradient, least squares method. *Num. Math.*, 51:209–227, 1987.

[10] M. Benantar and J. E. Flaherty. A Six color procedure for the parallel solution of Elliptic systems using the finite Quadtree structure. In J. Dongarra, P. Messina, D. C. Sorenson, and R. G. Voigt, editors, *Proceedings of the fourth SIAM conference on parallel processing for scientific computing*, pages 230–236, 1990.

[11] P. Concus, G. H. Golub, and D. P. O'Leary. A generalized conjugate gradient method for the numerical solution of elliptic partial differential equations. In James R. Bunch and Donald J. Rose, editors, *Sparse Matrix Computations*, pages 309–332, New York, 1976. Academic Press.

[12] I. S. Duff and G. A. Meurant. The effect of reordering on preconditioned conjugate gradients. *BIT*, 29:635–657, 1989.

[13] H. C. Elman. *Iterative Methods for Large Sparse Nonsymmetric Systems of Linear Equations*. PhD thesis, Yale University, Computer Science Dept., New Haven, CT., 1982.

[14] H. C. Elman and E. Agron. Ordering techniques for the precondiotioning conjugate gradient method on parallel computers. Technical Report UMIACS-TR-88-53, UMIACS, University of Maryland, College Park, MD, 1988.

[15] H. C. Elman and G. H. Golub. Iterative methods for cyclically reduced non-self-adjoint linear systems. Technical Report CS-TR-2145, Dept. of Computer Science, University of Maryland, College Park, MD, 1988.

[16] R. M. Ferencz. *Element-by-element preconditioning techniques for large scale vectorized finite element analysis in nonlinear solid and structural mechanics*. PhD thesis, Applied Mathematics, Stanford, CA, 1989.

[17] B. Fischer and R. Freund. Chebyshev polynomials are not always optimal. *J. Approximation Theory*, 65:–, 1991.

[18] R. Freund, M. H. Gutknecht, and N. M. Nachtigal. An implementation of the Look-Ahead Lanczos algorithm for non-Hermitian matrices, Part I. Technical Report 90-11, Massachusetts Institute of Technology, Cambridge, Massachusetts, 1990.

[19] R. Freund and N. M. Nachtigal. An implementation of the look-ahead lanczos algorithm for non-Hermitian matrices, Part II. Technical Report 90-11, Massachusetts Institute of Technology, Cambridge, Massachusetts, 1990.

[20] R. S. Varga G. H. Golub. Chebyshev semi iterative methods successive overre-

laxation iterative methods and second order Richardson iterative methods. *Numer. Math.*, 3:147–168, 1961.

[21] Alan George and Joseph W-H Liu. The evolution of the minimum degree ordering algorithm. *SIAM Review*, 31:1–19, 1989.

[22] A. L. Hageman and D. M. Young. *Applied Iterative Methods.* Academic Press, New York, 1981.

[23] T. J. R. Hughes, R. M. Ferencz, and J. O. Hallquist. Large-scale vectorized implicit calculations in solid mechanics on a cray x-mp/48 utilizing ebe preconditioning conjugate gradients. *Computer Methods in Applied Mechanics and Engineering*, 61:215–248, 1987.

[24] K. C. Jea and D. M. Young. Generalized conjugate gradient acceleration of non-symmetrizable iterative methods. *Linear Algebra Appl.*, 34:159–194, 1980.

[25] O. G. Johnson, C. A. Micchelli, and G. Paul. Polynomial preconditionings for conjugate gradient calculations. *SIAM J. Numer. Anal.*, 20:362–376, 1983.

[26] T. L. Jordan. Conjugate gradient preconditioners for vector and parallel processors. In G. N. Birkhoff and A. Schoenstadt, editors, *Elliptic problem solvers II, Proceedings of the elliptic problem solvers conference, Monterey CA. , Jan 10-12 1983*, pages 127–139. Academic Press, 1983.

[27] T. I. Karush, N. K. Madsen, and G. H. Rodrigue. Matrix multiplication by diagonals on vector/parallel processors. Technical Report UCUD, Lawrence Livermore National Lab., Livermore, CA, 1975.

[28] T. A. Manteuffel. The Tchebychev iteration for nonsymmetric linear systems. *Numer. Math.*, 28:307–327, 1977.

[29] T. A. Manteuffel. Adaptive procedure for estimation of parameter for the nonsymmetric Tchebychev iteration. *Numer. Math.*, 28:187–208, 1978.

[30] J. A. Meijerink and H. A. van der Vorst. An iterative solution method for linear systems of which the coefficient matrix is a symmetric M-matrix. *Math. Comp.*, 31(137):148–162, 1977.

[31] T. C. Oppe, W. Joubert, and D. R. Kincaid. Nspcg user's guide. a package for solving large linear systems by various iterative methods. Technical report, The University of Texas at Austin, 1988.

[32] T. C. Oppe and D. R. Kincaid. The performance of ITPACK on vector computers for solving large sparse linear systems arising in sample oil reservoir simulation problems. *Communications in applied numerical methods*, 2:1–7, 1986.

[33] J. M. Ortega. *Introduction to Parallel and Vector Solution of Linear Systems.* Plenum Press, New York, 1988.

[34] E. L Poole and J. M. Ortega. Mullticolor ICCG methods for vector computers. *SIAM J. Numer. Anal.*, 24:1394–1418, 1987.

[35] Y. Saad. Practical use of polynomial preconditionings for the conjugate gradient method. *SIAM J. Stat. Sci. Comput.*, 6:865–881, 1985.

[36] Y. Saad. Least squares polynomials in the complex plane and their use for solving sparse nonsymmetric linear systems. *SIAM J. Num. Anal.*, 24:155–169, 1987.

[37] Y. Saad. Krylov subspace methods on supercomputers. *SIAM J. Scient. Stat. Comput.*, 10:1200–1232, 1989.

[38] Y. Saad. SPARSKIT: A basic tool kit for sparse matrix computations. Technical Report 90-20, Research Institute for Advanced Computer Science, NASA Ames Research Center, Moffet Field, CA, 1990.

[39] Y. Saad. A flexible inner-outer preconditioned GMRES algorithm. Technical Report 91-279, Minnesota Supercomputer Institute, University of Minnesota, Minneapolis, Minnesota, 1991.

[40] Y. Saad. ILUT: a dual threshold incomplete ILU factorization. Technical report 92-38, Minnesota Supercomputer Institute, University of Minnesota, Minneapolis, 1992.

[41] Y. Saad and M. H. Schultz. Conjugate gradient-like algorithms for solving nonsymmetric linear systems. *Mathematics of Computation*, 44(170):417–424, 1985.

[42] Y. Saad and M. H. Schultz. GMRES: a generalized minimal residual algorithm for

solving nonsymmetric linear systems. *SIAM J. Sci. Statist. Comput.*, 7:856–869, 1986.

[43] J. H. Saltz. Automated problem scheduling and reduction of synchronization delay effects. Technical Report 87-22, ICASE, Hampton, VA, 1987.

[44] M. K. Seager. Parallelizing conjugate gradient for the CRAY X-MP. Technical report, Lawrence Livermore National Lab, Livermore, CA, 1984.

[45] F. Shakib. *Finite element analysis of the compressible Euler and Navier Stokes Equations.* PhD thesis, Aeronautics Dept., Stanford, CA, 1989.

[46] D. C. Smolarski and P. E. Saylor. An optimum iterative method for solving any linear system with a square matrix. *BIT*, 28:163–178, 1988.

[47] H. A. van der Vorst. High performance preconditioning. *SIAM j. Scient. Stat. Comput.*, 10:1174–1185, 1989.

[48] R. S. Varga. *Matrix Iterative Analysis.* Prentice Hall, Englewood Cliffs, NJ, 1962.

[49] V. Venkatakrishnan, H. D. Simon, and T. J. Barth. A MIMD Implementation of a Parallel Euler Solver for Unstructured Grids. Technical Report RNR-91-024, NASA Ames research center, Moffett Field, CA, 1991.

[50] P. K. W. Vinsome. Orthomin, an iterative method for solving sparse sets of simultaneous linear equations. In *Proceedings of the Fourth Symposium on Resevoir Simulation*, pages 149–159. Society of Petroleum Engineers of AIME, 1976.

[51] J. W. Watts-III. A conjugate gradient truncated direct method for the iterative solution of the reservoir simulation pressure equation. *Society of Petroleum Engineer Journal*, 21:345–353, 1981.

[52] O. Wing and J. W. Huang. A computation model of parallel solution of linear equations. *IEEE Transactions on Computers*, C-29:632–638, 1980.

[53] Y. S. Wong. Solving large elliptic difference equations on CYBER 205. *Parallel Comput.*, 6:195–207, 1988.

[54] C. H. Wu. A multicolour SOR method for the finite-element method. *J. of Comput. and App. Math.*, 30:283–294, 1990.

[55] D. M. Young. *Iterative solution of large linear systems.* Academic Press, New-York, 1971.

[56] D. M. Young, T. C. Oppe, D. R. Kincaid, and L. J. Hayes. On the use of vector computers for solving large sparse linear systems. Technical Report CNA-199, Center for Numerical Analysis, University of Texas at Austin, Austin, Texas, 1985.

[57] Z. Zlatev. Use of iterative refinement in the solution of sparse linear systems. *SIAM J. Numer. Anal.*, 19:381–399, 1982.

# A TWO-STAGE ITERATION FOR SOLVING NEARLY COMPLETELY DECOMPOSABLE MARKOV CHAINS

G.W. STEWART*, W.J. STEWART† AND D.F. McALLISTER†

**Abstract.** This paper is concerned with an iteration for determining the steady-state probability vector of a NCD Markov Chain. The states of these chains can be partitioned into aggregates with low probabilities of transitions between aggregates. The iteration consists of alternating block Gauss–Seidel iterations with Rayleigh–Ritz refinements. Under natural regularity conditions, the composite iteration reduces the error by a factor proportional to the size of the coupling between aggregates, so that the more loosely the chain is coupled, the faster the convergence.

**Key words.** Markov chain, NCD Chain, aggregation, Gauss–Seidel, Rayleigh–Ritz refinement

**1. Introduction.** Stochastic models of computer systems often result in Markov chains with a very large number of states. For example, a closed exponential queuing network with $K$ servers and $N$ customers may generate a chain that grows like $N^K/K!$ as $N$ becomes large. In order to determine the steady-state probability vector of a very large chain it is necessary to exploit any special structure that the underlying system may have.

In this paper we shall be concerned with nearly completely decomposable (NCD) Markov chains These chains model systems whose components can be clustered into loosely coupled aggregates of tightly interacting states. Mathematically, the states of such chains can be ordered so that the state transition probability matrix has the form

$$A = \begin{pmatrix} A_{11} & E_{12} & \cdots & E_{1l} \\ E_{21} & A_{22} & \cdots & E_{2l} \\ \vdots & \vdots & & \vdots \\ E_{l1} & E_{l2} & \cdots & A_{ll} \end{pmatrix}, \tag{1.1}$$

where the off-diagonal blocks $E_{ij}$ are small compared to the diagonal blocks $A_{ii}$. For later use we shall write

$$A = D + E, \tag{1.2}$$

where

$$D = \operatorname{diag}(A_{11}, A_{22}, \ldots, A_{ll}), \tag{1.3}$$

* Department of Computer Science, University of Maryland, College Park, MD 20742. This work was supported by the Air Force Office of Sponsored Research under Grant AFOSR-82-0078.

† Department of Computer Science, North Carolina State University, Raleigh, NC 27695-8206. This work was supported by the National Science Foundation under Grant MCS-83-00438.

and set

$$\epsilon = \|E\|,$$

where $\|\cdot\|$ denotes the spectral norm [8, Ch. 4].

Since $A$ is stochastic, it satisfies

$$Ax = x,$$

where $x$ denotes the vector whose components are all one. We shall assume that the chain is irreducible and aperiodic [3, Ch. 2], so that there is a unique, positive steady state vector $y$ that satisfies

$$y^{\mathrm{T}} A = y^{\mathrm{T}} \tag{1.4}$$

and

$$y^{\mathrm{T}} x = 1. \tag{1.5}$$

The purpose of this paper is to introduce a composite algorithm for computing $y$. The proposed algorithm is iterative, each iteration consisting of two parts — a sequence of *block Gauss–Seidel steps* followed by a *Rayleigh–Ritz refinement.* We shall now describe each step in detail.

The block Gauss–Seidel step is the iteration commonly used to solve the large linear systems arising in the numerical treatment of partial differential equations [12]. In this case we apply it to the system

$$y^{\mathrm{T}}(I - A) = 0, \tag{1.6}$$

which is equivalent to (1.4). Let $z$ denote the current approximation to $y$, and partition $z$ conformally with the partitioning of $A$ in (1.1):

$$z^{\mathrm{T}} = (z_1^{\mathrm{T}}\ z_2^{\mathrm{T}}\ \ldots\ z_l^{\mathrm{T}}). \tag{1.7}$$

The block Gauss–Seidel iteration produces a new approximation $\tilde{z}$ to $y$ according to the formula

$$\tilde{z}_j^{\mathrm{T}} = \Big(\sum_{i<j} \tilde{z}_i^{\mathrm{T}} E_{ij} + \sum_{i>j} z_i^{\mathrm{T}} E_{ij}\Big)(I - A_{jj})^{-1} \qquad (j = 1, 2, \ldots, l). \tag{1.8}$$

In other words $\tilde{z}_j$ is just the vector obtained by solving the $j$th row of the partitioned form of (1.6) while holding $\tilde{z}_1 \ldots \tilde{z}_{j-1}$ and $z_{j+1} \ldots z_l$ fixed.

The Rayleigh–Ritz step is a standard method for extracting an approximate eigenvector from a subspace that nearly contains the eigenvector. Again let $z$, the current approximation to $y$, be partitioned as in (1.7), and assume that the components of $z$ are positive. Let $\hat{z}_j = z_j/\|z_j\|_1$, where $\|\cdot\|_1$ denotes the 1-norm [8, Ch. 4]. The space from which the new approximation $\tilde{z}$ is to be extracted is spanned by the columns of the matrix

$$T_z = \begin{pmatrix} \hat{z}_1 & 0 & \cdots & 0 \\ 0 & \hat{z}_2 & \cdots & 0 \\ \vdots & \vdots & & \vdots \\ 0 & 0 & \cdots & \hat{z}_l \end{pmatrix}. \tag{1.9}$$

The algorithm goes as follows. Set

$$S = \begin{pmatrix} x_1 & 0 & \cdots & 0 \\ 0 & x_2 & \cdots & 0 \\ \vdots & \vdots & & \vdots \\ 0 & 0 & \cdots & x_l \end{pmatrix},$$

where the $x_i$ are from a partition of $x$ conformal with (1.1). Compute

$$B_z = T_z^{\mathrm{T}} AS. \tag{1.10}$$

It is easily verified that $B_z$ is an irreducible, aperiodic, stochastic matrix and therefore has a unique, positive eigenvector $v_z$ that satisfies

$$v_z^{\mathrm{T}} B_z = v_z^{\mathrm{T}} \qquad (\|v_z\|_1 = 1).$$

The new approximation $\tilde{z}$ to $y$ is given by

$$\tilde{z} = T_z v_z.$$

The iteration proposed here consists of alternating sequences of Gauss–Seidel and Rayleigh–Ritz steps. In the next two sections of this paper we shall give a mathematical analysis of this method. However, the treatment is quite involved, and it is appropriate to summarize our results here. Those who are uninterested in the supporting details should pass directly from the end of this section to §4, where implementation issues are discussed.

Provided certain regularity conditions are satisfied, our results hold for all sufficiently small $\epsilon$. In Section 2 it will be shown that the vector $z$ can be written in the form

$$z = y + Y_2 g_2 + Y_3 g_3, \tag{1.11}$$

where $Y_2$ is an $n \times (l-1)$ matrix and $Y_3$ is an $n \times (n-l)$ matrix, both constructed from the matrix $A$. If we set

$$\gamma_i = \|g_i\| \qquad (i = 2, 3), \tag{1.12}$$

then the sizes of the $\gamma_i$ tell how good an approximation $z$ is to $y$. It turns out that the Gauss–Seidel step reduces $\gamma_3$ by a factor of order $\epsilon$. The subsequent Rayleigh–Ritz step reduces $\gamma_2$, again by a factor of order $\epsilon$. Thus the composite iteration reduces the error by a factor of order $\epsilon$. Both steps are necessary, since the Gauss–Seidel step may reduce $\gamma_2$ only slowly, while the Rayleigh–Ritz step is stationary after one application (because $T_z = T_{\hat{z}}$).

In this paper we shall analyze only the Gauss–Seidel step, since the Rayleigh–Ritz step has been analyzed elsewhere [4]. The latter is closely related to methods of aggregation, which were first proposed in this connection by Simon and Ando [5]. The idea is to determine a vector $z$ such that $T_z$ is a good approximation to $T_y$ and then apply a Rayleigh–Ritz step to approximate $y$. The diagonals of $T_z$ are usually found by solving eigenvalue problems associated with with the $A_{ii}$. For example, Stewart [7] takes the $\hat{z}_i$ to be the normalized left eigenvectors of the $A_{ii}$; Courtois [1] takes them to be the normalized left eigenvectors of stochastic approximations $A_{ii}^*$ to the $A_{ii}$. Under appropriate assumptions, either method produces an $O(\epsilon)$ approximation to $y$.

The drawback of this form of aggregation is that it produces a single result, which may or may not be accurate enough. To remedy this defect, two workers have proposed composite iterative schemes that reduce the error by a factor of order $\epsilon$. Takahashi [10] has proposed a method that turns out to be very close to the one analyzed here; so close, in fact, that the analysis of our method may be easily extended to prove the convergence of his.

The method of Vantilborgh [11] is quite different in character. He uses an approximation to $y$ to construct stochastic matrices $A_{ii}^*$, each with the property that its steady-state vector $\hat{z}_i$ is an improved approximation to the corresponding part of $y$, suitably normalized. These vectors are used in a Rayleigh–Ritz step, as described above. Vantilborgh indicates that when this process is iterated, it provides an $O(\epsilon)$ reduction in the error for each step. The chief drawback of the algorithm is that it requires the solution of $l$ eigenvalue problems to compute the $\hat{z}_i$.

The paper is organized as follows. In the next section material from [4] on the structure of NCD Markov chains is reviewed. Careful attention will be devoted to the conditions that must obtain as $\epsilon$ approaches zero in order for our results to be valid. In §3 the Gauss–Seidel step is analyzed. The paper concludes with a section containing a discussion of the computational issues and an example.

**2. The Structure of NCD Markov Chains.** In this section we shall review some results found in [4] on the structure of NCD Markov chains. The purpose is twofold. First, among the results is a precise statement of the behavior of the Rayleigh–Ritz refinement. Second, the notation and results are needed to establish the properties of the Gauss–Seidel step.

We wish for our results to hold uniformly as $\epsilon$ approaches zero; i.e., as $A$ approaches a block diagonal matrix. This requires that we restrict the way in which $A$ approaches block diagonality; for very simple counter examples show that a small $\epsilon$ is not by itself sufficient to guarantee the structure usually associated with NCD matrices. One approach is to assume a specific parameterization for $A$. For example, in [1] and [11] it is assumed that $A$ can be written in the form

$$A = A^* + \epsilon B,$$

where $A^*$ is a block diagonal stochastic matrix. Although this approach works, it has two drawbacks. First, it does not allow the elements of $E$ to approach zero other than linearly. Second, it presupposes that the approach to the limit is along the fixed path determined by the matrix B. We therefore prefer to place different restrictions on the behavior of $A$ as $\epsilon$ approaches zero, restrictions which at once overcome the above drawbacks and give insight into how the theory can fail.

The first condition specifies how the components of $y$ behave. Write

$$y^{\mathrm{T}} = (y_1^{\mathrm{T}}\ y_2^{\mathrm{T}}\ \cdots\ y_l^{\mathrm{T}}),$$

where the partitioning is conformal with the underlying partition of the chain. Then we require the following for our first condition.[1]

**Regularity Condition 1.** There is a constant $m_1 > 0$ such that

$$\|y_i\|_1 \geq m_1 \qquad (i = 1, 2, \ldots, l). \tag{2.1}$$

This condition is related to the asymptotic block irreducibility of $A$. Specifically, set $\hat{y}_i = y_i/\|y_i\|_1$ and let $B_y$ be defined in analogy with (1.10). Then the diagonal elements $\hat{y}_i^{\mathrm{T}} A_{ii} x_i$ of $B_y$ approach one, while the off-diagonal elements $\hat{y}_i^{\mathrm{T}} E_{ij} x_j$ approach zero at least as fast as $\epsilon$. If some of the off-diagonal elements of $B_y$ were to approach zero faster than others in such a way that $B_y$ approaches reducibility, then one would expect some of the components of the left eigenvector $v_y$ of $B_y$ to approach zero. But it is easily verified that

$$v_y^{\mathrm{T}} = (\|y_1\|_1, \|y_2\|_1, \ldots, \|y_l\|_1). \tag{2.2}$$

Thus (2.1) guards against the effects of asymptotic reducibility.

The second condition requires some further notation. In [4] it is shown that there are matrices $J_i$ and $K_i$ such that

$$(x_i\ J_i)^{-1} = (\hat{y}_i\ K_i)^{\mathrm{T}} \qquad (i = 1, 2, \ldots, l). \tag{2.3}$$

Moreover, for $i = 1, 2, \ldots, l$ the norms of the $J_i$ and $K_i$ are bounded by quantities that are independent of $\epsilon$. Let

$$(\hat{y}_i\ K_i)^{\mathrm{T}} A_{ii} (x_i\ J_i) = \begin{pmatrix} \beta_i & h_i^{\mathrm{T}} \\ g_i & C_i \end{pmatrix} \qquad (i = 1, 2, \ldots, l). \tag{2.4}$$

[1] Subsequent to the writing of this paper, X. Sun [9] has shown that the first condition is unnecessary.

Our second regularity condition may now be stated as follows.

**Regularity Condition 2.** There is a constant $M_2$ such that

$$\|(I - C_i)^{-1}\| < M_2 \qquad (i = 1, 2, \ldots, l). \tag{2.5}$$

This condition circumscribes the behavior of the smallest $n-l$ eigenvalues of $A$. To see this, note that as $\epsilon$ approaches zero, the eigenvalues of $A$ approach those of the $A_{ii}$, or equivalently those of the matrices (2.4), which by (2.3) are similar to the $A_{ii}$. Now it is shown in [4] that $g$ and $h$ approach zero as $\epsilon$ approaches zero, and it is easy to verify that the $\beta_i$ approach one. Thus the smallest $n - l$ eigenvalues of $A$ approach the eigenvalues of the $C_i$. Since the norm of a matrix is a bound on its eigenvalues, if $c$ denotes any eigenvalue of $C_i$ then it follows from (2.5) that $|1 - c|^{-1} < M_2$, or equivalently

$$|1 - c| > \frac{1}{M_2}.$$

Thus the second regularity condition insures that the smallest $n - l$ eigenvalues of $A$ remain uniformly bounded away from one.

In order to describe the structure of NCD Markov chains that satisfy the two regularity conditions, we must introduce still more notation. We have observed that the vector $v_y$ defined by (2.2) is a left eigenvector of $B_y$. Similarly the vector $u = (1, 1, \ldots, 1)^{\mathrm{T}}$ is a right eigenvector of $B_y$. By the remarks surrounding equation (2.3), we can find uniformly bounded matrices $U$ and $V$ of dimension $l \times (l - 1)$ such that

$$(u\ U)^{-1} = (v_y\ V)^{\mathrm{T}}.$$

Because $u$ and $v_y$ are right and left eigenvectors of $B_y$,

$$(v_y\ V)^{\mathrm{T}} B_y (u\ U) = \begin{pmatrix} 1 & 0 \\ 0 & \tilde{A}_2 \end{pmatrix}; \tag{2.6}$$

that is the similarity transformation (2.6) block diagonalizes $B_y$, placing the eigenvalue one in the upper left and the remaining eigenvalues in the matrix $\tilde{A}_2$. Finally set

$$J = \begin{pmatrix} J_1 & 0 & \cdots & 0 \\ 0 & J_2 & \cdots & 0 \\ \vdots & \vdots & & \vdots \\ 0 & 0 & \cdots & J_l \end{pmatrix},$$

$$K = \begin{pmatrix} K_1 & 0 & \cdots & 0 \\ 0 & K_2 & \cdots & 0 \\ \vdots & \vdots & & \vdots \\ 0 & 0 & \cdots & K_l \end{pmatrix},$$

and

$$\tilde{A}_3 = K^{\mathrm{T}} A J. \tag{2.7}$$

We are now in a position to state the main result of this section, which asserts the existence and describes the structure of matrices $(x\ X_2\ X_3)$ and $(y\ Y_2\ Y_3)$ satisfying

$$(x\ X_2\ X_3)^{-1} = (y\ Y_2\ Y_3)^{\mathrm{T}} \tag{2.8}$$

and

$$(y\ Y_2\ Y_3)^{\mathrm{T}} A (x\ X_2\ X_3) = \operatorname{diag}(1, A_2, A_3). \tag{2.9}$$

Here $A_2$ is of order $l-1$ and $A_3$ is of order $n-l$. The proof may be found in [4].

THEOREM 2.1. *Let $A$ satisfy Regularity Conditions 1 and 2. Then for all sufficiently small $\epsilon$ the decomposition* (2.8–2.9) *exists. Moreover, there are matrices $P$ and $Q$ satisfying*

$$\|P\|, \|Q\| = O(\epsilon) \tag{2.10}$$

*such that*

$$\begin{array}{ll} 1.\ x = Su, & 2.\ y = T_y v_y, \\ 3.\ X_2 = SU + JP + O(\epsilon^2), & 4.\ Y_2 = T_y V + KQ, \\ 5.\ X_3 = J - SUQ^{\mathrm{T}} + O(\epsilon^2), & 6.\ Y_3 = K - T_y V P^{\mathrm{T}}, \\ 7.\ A_2 = \tilde{A}_2 + O(\epsilon^2), & 8.\ A_3 = \tilde{A}_3 + O(\epsilon^2). \end{array} \tag{2.11}$$

There are several comments to be made about this theorem. In the first place, it follows from (2.9) that

$$Y_i^{\mathrm{T}} A = A_i Y_i^{\mathrm{T}} \qquad (i = 1, 2).$$

This shows that the space spanned by the columns of $Y_i$ is a left invariant subspace of $A$; that is, it is unchanged when it is transformed by $A^{\mathrm{T}}$. The matrix $A_i$ is the representation of $A$ restricted to the column space of $Y_i$.

Since from (2.11) the powers of $A$ satisfy

$$Y_i^{\mathrm{T}} A^s = A_i^s Y_i^{\mathrm{T}} \qquad (i = 1, 2),$$

the eigenvalues of $A_i$ control the behavior of the Markov chain on the space spanned by $Y_i$. The asymptotic behavior of these eigenvalues is easily determined. From (1.5), (2.6) and (2.11) it is seen that

$$A_2 = I + O(\epsilon). \tag{2.12}$$

Consequently the eigenvalues of $A_2$ are near one, and as $s$ becomes large $A_2^s$ approaches zero slowly; that is, $Y_2$ corresponds to a slow transient of the Markov chain. On the other hand, from (2.4), (2.7) and (2.11) it is seen that

$$A_3 = \mathrm{diag}(C_1, C_2, \ldots, C_l) + O(\epsilon). \tag{2.13}$$

From Regularity Condition 2, the eigenvalues of the $C_i$ are bounded away from one. Consequently, $A_3^s$ approaches zero more swiftly than $A_2^s$; that is, $Y_3$ corresponds to a fast transient. This behavior of NCD chains was first noted by Simon and Ando [5]. [We note in passing that a condition different from Regularity Condition 2 was used by the authors of this paper to establish the results in [4]. However, that condition is implied by Regularity Condition 2 along with (2.12) and (2.13).]

The spaces associated with the slow and fast transients are highly structured. For example, (2.11.4) shows that the space corresponding to the slow transient is essentially a subspace of space spanned by $T_y$; i.e., it is made up of pieces of the eigenvector $y$. This is why the Rayleigh–Ritz technique is so good at eliminating errors in the slow transient space; if the $\hat{z}_i$ are good approximations to the $\hat{y}_i$ [see (1.9)], then $T_z$ contains a good approximation to the slow transient error and the Rayleigh–Ritz process uses it to purge the error.

On the other hand, the space associated with the fast transient is essentially spanned by $K$. This fact will be crucial in explaining why the Gauss–Seidel step tends to purge errors along the fast transient.

We turn now to the properties of the Rayleigh–Ritz step. The basic result requires yet another regularity condition, designed to keep the slow transient apart from the steady state. This can be done by keeping the eigenvalues of $\tilde{A}_2$ in (2.6) bounded away from one by a quantity proportional to $\epsilon$. Rather than work directly with eigenvalues, we work with the norm of an appropriate matrix, as we did in Regularity Condition 2.

**Regularity Condition 3.** There is a constant $M_3$ such that

$$\|(I - \tilde{A}_2)^{-1}\| < M_3\epsilon^{-1}$$

With the third regularity condition added to the first two, the following theorem is valid [4].

THEOREM 2.2. *Let $A$ satisfy Regularity Conditions 1, 2, and 3. Let $z > 0$ be given with $\|z\|_1 = 1$. Let $\gamma_i$ be defined by (1.12). Let $\tilde{z}$ be the vector produced by one step of the Rayleigh–Ritz step and let $\tilde{\gamma}_i$ be defined in analogy with (1.12). Then for all sufficiently small $\epsilon$, $\gamma_2$ and $\gamma_3$ there are constants $N_2, N_3 > 0$ such that*

$$\tilde{\gamma}_i <= \epsilon N_2 \gamma_2 + N_3 \gamma_3 \qquad (i = 1, 2). \tag{2.14}$$

From this theorem it is seen that the Rayleigh–Ritz step is a $\gamma_2$ reducer. However, it is only effective when $\gamma_3$ is less than $\gamma_2$. Thus it must be used in conjunction with a $\gamma_3$ reducer. In the next section we shall show that the Gauss–Seidel step has this property.

**3. Analysis of the Gauss–Seidel Step.** In this section we shall be concerned with the analysis of the Gauss–Seidel step defined by (1.8). Although the details of the analysis are tedious, the underlying idea is straightforward. Let $z > 0$ have the expansion (1.11). Let $\tilde{z}$ be determined by (1.8) and write

$$\tilde{z} = \tilde{\gamma}_1 y + Y_2 \tilde{g}_2 + Y_3 \tilde{g}_3.$$

From (2.8) it follows that

$$\tilde{\gamma}_1 = \tilde{z}^{\mathrm{T}} x, \tag{3.1}$$

and

$$\tilde{g}_i^{\mathrm{T}} = \tilde{z}^{\mathrm{T}} X_i \qquad (i = 2, 3). \tag{3.2}$$

Thus the problem is to develop expressions for the products on the right hand sides of (3.1) and (3.2).

The first step is to express $\tilde{z}$ in matrix form. Referring to (1.2), let

$$A = D + L + U,$$

where $L$ is strictly lower triangular and $U$ is strictly upper triangular. After some manipulation, $\tilde{z}$ can be written in the form

$$\tilde{z}^{\mathrm{T}} = z^{\mathrm{T}} L[I - (I - D)^{-1} U]^{-1} (I - D)^{-1}. \tag{3.3}$$

If the expansion (1.11) for $z$ is substituted into (3.3), the result is

$$\tilde{z}^{\mathrm{T}} = y^{\mathrm{T}} + g_2^{\mathrm{T}} Y_2^{\mathrm{T}} Z (I - D)^{-1} + g_3^{\mathrm{T}} Y_3^{\mathrm{T}} Z (I - D)^{-1}, \tag{3.4}$$

where

$$Z = L[I - (I - D)^{-1} U]^{-1}$$

(the simple form of the first term in the right-hand side of (3.4) is due to the fact that $y$ is a fixed point of the Gauss–Seidel iteration).

The key to the analysis is to establish the behavior of the quantities in (3.4), which is done in the lemma below. The results may be summarized by saying that as $\epsilon$ approaches zero 1. $(I - D)^{-1}$ becomes large, but not too fast; 2. $(I - D)^{-1}$ remains bounded in the column space of $X_3$; 3. $Z$ approaches zero as fast as $(I - D)^{-1}$ becomes large. Specifically,

LEMMA 3.1. *Let $A$ satisfy the three regularity conditions of Section 2. Then*

$$\begin{aligned} &1.\ \|(I - D)^{-1}\| = O(\epsilon^{-1}) \\ &2.\ \|(I - D)^{-1} X_3\| = O(1) \\ &3.\ \|Z\| = O(\epsilon) \end{aligned} \tag{3.5}$$

**Proof.** From (1.3), to establish (3.5.1) it is sufficient to establish that $\|(I - A_{ii})^{-1}\| = O(\epsilon^{-1})$. In view of (2.4) and the uniform boundedness of the $J_i$ and the $K_i$, it is sufficient to establish that

$$\begin{pmatrix} 1-\beta_i & -h_i^{\mathrm{T}} \\ -g_i & I-C_i \end{pmatrix} = O(\epsilon^{-1}).$$

Now

$$\begin{pmatrix} 1-\beta_i & -h_i^{\mathrm{T}} \\ -g_i & I-C_i \end{pmatrix} = \begin{pmatrix} 1 & 0 \\ -(1-\beta_i)^{-1}g_i & I \end{pmatrix} \begin{pmatrix} 1-\beta_i & -h_i^{\mathrm{T}} \\ 0 & I-C_i-(1-\beta_i)^{-1}g_ih_i^{\mathrm{T}} \end{pmatrix}, \tag{3.6}$$

which can be verified by multiplying the factors on the right hand side. In the appendix we shall show that

$$(1-\beta_i)^{-1} = O(\epsilon^{-1}). \tag{3.7}$$

Since $\|g_i\|$ and $\|h_i\|$ are of order $\epsilon$, it follows from (3.6) and (3.7) that

$$\begin{pmatrix} 1-\beta_i & -h_i^{\mathrm{T}} \\ -g_i & I-C_i \end{pmatrix} = \begin{pmatrix} 1 & 0 \\ O(1) & I \end{pmatrix} \begin{pmatrix} 1-\beta_i & O(\epsilon) \\ 0 & I-C_i+O(\epsilon) \end{pmatrix}, \tag{3.8}$$

From (2.5) and standard perturbation theory for matrix inverses [8, Ch. 4], $[I - C_i + O((\epsilon))]^{-1} = O(1)$. Hence from (3.8)

$$\begin{pmatrix} 1-\beta_i & -h_i^{\mathrm{T}} \\ -g_i & I-C_i \end{pmatrix}^{-1} = \begin{pmatrix} O(\epsilon^{-1}) & O(1) \\ O(1) & O(1) \end{pmatrix}, \tag{3.9}$$

which establishes (3.5.1).

To establish (3.5.2) note that by (2.11.5) it is sufficient to show that $(I - D)^{-1}J = O(1)$, or in view of the block diagonality of $D$ and $K$ that $(I - A_{ii})^{-1}J_i = O(1)$. From (2.4)

$$(I - A_{ii})^{-1}J_i = (x_i\ J_i)\begin{pmatrix} 1-\beta_i & -h_i^{\mathrm{T}} \\ -g_i & I-C_i \end{pmatrix}^{-1}(x_i\ J_i)^{-1}J_i.$$

Hence from (3.9)

$$(I - A_{ii})^{-1}J_i = (x_i\ J_i)\begin{pmatrix} O(\epsilon^{-1}) & O(1) \\ O(1) & O(1) \end{pmatrix}\begin{pmatrix} 0 \\ 1 \end{pmatrix} = (x_i\ J_i)\cdot O(1),$$

and the result follows from the fact that $(x_i\ J_i)$ is uniformly bounded.

To establish (3.5.3) note that since $L = O(\epsilon)$ it is sufficient to show that

$$[I - (I - D)^{-1}U]^{-1} = O(1).$$

Set $\bar{U} = (I-D)^{-1}U$. Since $U = O(\epsilon)$, it follows from (3.5.1) that $\bar{U} = O(1)$. Since $\bar{U}$ is strictly lower triangular, the geometric series

$$(I - \bar{U})^{-1} = I + \bar{U} + \bar{U}^2 + \cdots$$

terminates. Thus $(I - \bar{U})^{-1} = O(1)$ which establishes (3.5.3) and the lemma.

With Lemma 3.1 established, the analysis of the Gauss–Seidel step amounts to little more than postmultiplying (3.4) by $x$, $X_2$ and $X_3$ and applying the lemma. For example,

$$\begin{aligned}\tilde{\gamma}_3 &= \|\tilde{z}^{\mathrm{T}} X_3\| \\ &= \|g_2\|\|Y_2^{\mathrm{T}} Z\|\|(I-D)^{-1}X_3\| + \|g_3\|\|Y_2^{\mathrm{T}} Z\|\|(I-D)^{-1}X_3\|.\end{aligned}$$

Hence from (3.5.2) and (3.5.3) it follows that there is a constant $L^3$ such that $\tilde{\gamma}_3 <= \epsilon L^3(\gamma_2 + \gamma_3)$. The other bounds are obtained similarly.

THEOREM 3.2. *In the notation introduced above, if $A$ satisfies the three regularity conditions, then there are constants $L^i$ $(i = 1, 2, 3)$ such that*

$$\begin{aligned}&a.\ |1 - \tilde{\gamma}_1| \le L^1(\gamma_2 + \gamma_3),\\ &b.\ \tilde{\gamma}_2 \le L^2(\gamma_2 + \gamma_3),\\ &c.\ \tilde{\gamma}_3 \le \epsilon L^3(\gamma_2 + \gamma_3).\end{aligned}$$

Theorem 3.2 allows us to give a local convergence proof of the two step iteration. First assume that initially

$$L^1(\gamma_2 + \gamma_3) \le \frac{1}{2}, \tag{3.10}$$

so that when the vector $\tilde{z}$ obtained from the Gauss–Seidel step is normalized to give a vector $\bar{z}$, the resulting $\bar{\gamma}_i$ satisfy

$$\begin{aligned}&b.\ \bar{\gamma}_2 \le 2L^2(\gamma_2 + \gamma_3)\\ &c.\ \bar{\gamma}_3 \le 2\epsilon L^3(\gamma_2 + \gamma_3)\end{aligned} \tag{3.11}$$

In matrix notation (3.11) becomes

$$\begin{pmatrix} \bar{\gamma}_2 \\ \bar{\gamma}_3 \end{pmatrix} \le 2 \begin{pmatrix} L^2 & L^2 \\ \epsilon L^3 & \epsilon L^3 \end{pmatrix} \begin{pmatrix} \gamma_2 \\ \gamma_3 \end{pmatrix}.$$

If $\hat{z}$ denotes the result of applying a Rayleigh–Ritz refinement to $\bar{z}$, then from (2.14) it follows that

$$\begin{aligned}\begin{pmatrix} \bar{\gamma}_2 \\ \bar{\gamma}_3 \end{pmatrix} &\le 2 \begin{pmatrix} \epsilon N_2 & N_3 \\ \epsilon N_2 & N_3 \end{pmatrix} \begin{pmatrix} L^2 & L^2 \\ \epsilon L^3 & \epsilon L^3 \end{pmatrix} \begin{pmatrix} \gamma_2 \\ \gamma_3 \end{pmatrix} \\ &= 2\epsilon(N_2 L^2 + N^3 L^3) \begin{pmatrix} 1 & 1 \\ 1 & 1 \end{pmatrix} \begin{pmatrix} \gamma_2 \\ \gamma_3 \end{pmatrix}\end{aligned}$$

It follows that

$$\hat{\gamma}_2 + \hat{\gamma}_3 \le 4\epsilon(N_2L^2 + N_3L^3)(\gamma_2 + \gamma_3).$$

Thus if

$$4\epsilon(N_2L^2 + N_3L^3) < 1,$$

then (3.10) remains satisfied, and the iteration converges. In other words the composite iteration consisting of alternating Gauss–Seidel and Rayleigh–Ritz steps converges provided $\epsilon$ is sufficiently small and the starting approximation is sufficiently good.

There are three comments to be made about these results. First, although we have applied Theorem 3.2 to analyze the case where a Rayleigh–Ritz step is alternated with a Gauss-Siedel step, in practice it may be preferable to perform two or more Gauss–Seidel steps. The best mix will depend on factors, such as the observed rates of convergence and the relative cost of each kind of step, which are best evaluated in the course of the iteration. We shall give some examples in the next section.

The second observation is that the analysis extends with little modification to the method of Takahashi [10]. His method differs from ours in the way intermediate quantities are scaled in the Gauss–Seidel step. Specifically, if we write

$$W_z = \operatorname{diag}(\|z_1\|_1 I_{n_1}, \ldots, \|z_l\|_1 I_{n_l}),$$

then the iterate $\tilde{z}$ produced by Takahashi's method satisfies

$$\tilde{z}^{\mathrm{T}} = z^{\mathrm{T}} L\big(I - (I-D)^{-1} W_{\tilde{z}}^{-1} W_z U\big)^{-1} (I-D)^{-1}.$$

It is easily shown that as $z \to y$, the matrix $W_{\tilde{z}}^{-1}W_z \to I$. Consequently, if $\epsilon$ is small enough and $z$ is near enough $y$, the two methods generate almost identical iterates, both of which reduce the error by a factor uniformly less than one.

Finally we note that although the bounds derived above are sufficient to establish the convergence of the two-stage iteration, they by no means give an adequate description of how the iteration behaves in practice. For it turns out that the Gauss–Seidel step, in addition to being a $\gamma_3$ reducer, is also a $\gamma_2$ reducer. How this comes about is an open research problem.

**4. Implementation and Examples.** An important advantage of the two-stage iteration is that each of its steps is comparatively easy to implement. The Gauss–Seidel step (1.8) requires that the system of equations

$$\tilde{z}_j^{\mathrm{T}}(I - A_{jj}) = \Big(\sum_{i<j} \tilde{z}_i^{\mathrm{T}} E_{ij} + \sum_{i>j} z_i^{\mathrm{T}} E_{ij}\Big)$$

be solved for $j = 1, 2, \ldots, l$. The preferred method is to compute an LU factorization of the matrix $(I - A_{jj})$ and use this factorization to solve the system [2]. Because the matrix does not change from iteration to iteration, the factorization need be done only once, which represents a great savings in computational effort. In fact, the floating-point operation count is approximately the cost of a single matrix-vector multiplication: $n^2$ flops,[2] less if advantage is taken of the sparsity of the matrix.

In the Rayleigh–Ritz step, the formation of $B_z$ will require approximately the same amount of work as one Gauss–Seidel step. Since only the dominant left eigenvector of $B_z$ is required and the corresponding eigenvalue is known, it should be computed by the inverse power method [8, Ch. 7]. This requires that an LU decomposition of $I - B_z$ be calculated at a cost of $l^3/3$ flops. In large problems this calculation will be insignificant compared to the computation of $B_z$, so that the Gauss–Seidel step and the Rayleigh–Ritz step will represent comparable computations.

To illustrate the procedure we consider two examples, based on a model investigated by Vantilborgh in [11], where further references may be found. The following is Vantilborgh's description of the model.

> A finite number $N$ of active user terminals generates random requests for program execution. User programs are executed on a multiprogrammed basis by a CPU/Main Memory server $S_1$ and a paged secondary-memory server $S_2$. A program is admitted to the multiprogramming mix if less than $M$ programs are jointly present at $S_1$ and $S_2$; programs not admitted to this mix are kept waiting in a queue $Q_3$. At the end of a service, a program generates a disk request with probability $\delta$, requires another quantum of execution with probability $\phi$, or terminates. We use $i_1$ and $i_2$ to denote the congestion at $S_1$ and $S_2$.
>
> We assume
>
> (i) The think time of a user has a negative exponential distribution with parameter $\lambda$; [Throughout our examples we choose $\lambda = 0.001$.]
>
> (ii) $S_1$ and $S_2$ have exponentially distributed service times with parameters $\mu_1(i_1)$ and $\mu_2(i_2)$; [As in Vantilborgh, we choose the specific values $\mu_1(i_1) = 64/(i_1 + 16)$ and $\mu_2(i_2) = 3i_2/(i_2 + 6)$.]
>
> (iii) $\phi$, and $\delta$ are independent of the state of the network; [We choose $\phi = 0.03$ and $\delta = 0.95$.]

In our first experiment with this model, we choose values of $N(6)$ and $M(3)$ so that the resulting probability matrix is small enough (order 22) to analyze completely using the QR algorithm [6]. This permits us to compute the norms $\gamma_1$ and $\gamma_2$ of the errors along the slow and fast transient subspaces. Table 4.1 presents the results first of alternating Rayleigh–Ritz steps with Gauss–Seidel steps and then of following a Rayleigh–Ritz step with five Gauss–Seidel steps. In both cases the starting vector was

[2] A flop is one floating-point addition and one floating-point multiplication, usually accompanied by two array accesses.

TABLE 4.1
*Vantliborgh Model*

| $N = 6$ $M = 3$ | | | | | |
|---|---|---|---|---|---|
| step | $\gamma_2$ | $\gamma_3$ | step | $\gamma_2$ | $\gamma_3$ |
| I | 4.2E+00 | 7.2E+00 | I | 4.2E+00 | 7.2E+00 |
| RR | 5.8E−03 | 5.5E−03 | RR | 5.8E−03 | 5.5E−03 |
| GS | 2.5E−03 | 6.5E−05 | GS | 2.5E−03 | 6.5E−05 |
| RR | 3.9E−05 | 3.6E−04 | GS | 4.9E−04 | 1.4E−07 |
| GS | 1.6E−05 | 2.5E−07 | GS | 1.3E−04 | 3.3E−08 |
| RR | 2.9E−07 | 2.6E−06 | GS | 3.6E−05 | 9.7E−09 |
| GS | 1.2E−07 | 1.6E−09 | GS | 1.1E−05 | 2.9E−09 |
| RR | 2.1E−09 | 1.9E−08 | RR | 1.5E−07 | 1.4E−06 |
| GS | 8.4E−10 | 1.1E−11 | GS | 6.1E−08 | 9.3E−10 |
| RR | 1.5E−11 | 1.3E−10 | GS | 1.8E−08 | 3.3E−12 |
| GS | 6.0E−12 | 7.6E−14 | GS | 4.5E−09 | 1.1E−12 |
| RR | 1.3E−13 | 9.4E−13 | GS | 1.2E−08 | 3.2E−13 |
| GS | 8.8E−14 | 8.0E−16 | GS | 3.5E−10 | 9.5E−14 |
| | | | RR | 5.6E−12 | 5.0E−11 |
| | | | GS | 2.2E−12 | 3.4E−14 |
| | | | GS | 6.6E−13 | 3.0E−16 |
| | | | GS | 1.9E−13 | 3.0E−16 |
| | | | GS | 9.0E−14 | 3.0E−16 |
| | | | GS | 6.6E−14 | 3.0E−16 |

the vector of all ones. In the column labeled "step", the I indicates the initial values of $\gamma_2$ and $\gamma_3$, RR indicates the values after a Rayleigh–Ritz step, and GS the values after a Gauss–Seidel step. It is seen from this table, that increasing the number of Gauss–Seidel steps does not improve the convergence (experiments not summarized in the table show that an intermediate number of Gauss–Seidel steps does not help either).

The results also show that the Gauss–Seidel step reduces $\gamma_2$, although not as dramatically as a well-timed Rayleigh–Ritz step. This phenomena, which was alluded to in the last section, represents a gap in our understanding of the composite algorithm.

For the second example we choose $N = 50$ and $M = 30$, which gives a matrix of order 1,116 with diagonal blocks of order 1, 2, 3, ..., 29, 30, 31, 31, ..., 31. The remaining parameters were left unaltered. Since the matrix is too large to analyze into invariant subspaces, we cannot compute $\gamma_2$ and $\gamma_3$. Instead we compute the error in the steady-state vector, which was obtained by iterating the two-stage iteration to convergence.

Table 4.2 shows the number of iterations required to achieve a given accuracy for a Rayleigh–Ritz step followed by a fixed number of Gauss–Seidel steps. It is seen that the best behavior is achieved when each Rayleigh–Ritz step is followed by three Gauss–Seidel steps.

TABLE 4.2
*Vantliborgh Model*

| $N = 50$ $M = 30$<br>No. of Steps to Attain<br>No. a Given Accuracy | | | | | |
|---|---|---|---|---|---|
| Decimal Places | Number of G–S Steps 1 | 2 | 3 | 4 | 5 |
| 2 | 5 | 4 | 5 | 6 | 7 |
| 3 | 7 | | | | |
| 4 | 9 | 7 | 9 | 11 | 13 |
| 5 | 12 | 10 | | | |
| 6 | 16 | 13 | 13 | 16 | 19 |
| 7 | 20 | 16 | | | |
| 8 | 25 | 19 | 17 | 21 | 25 |
| 9 | 30 | 23 | 21 | | |
| 10 | 36 | 27 | 25 | 26 | 31 |
| 11 | 42 | 31 | 29 | 31 | |
| 12 | 52 | 38 | 33 | 36 | 37 |
| 13 | 60 | 44 | 38 | 38 | 43 |

**Appendix: Proof of Equation (3.7).** Let

$$X = \begin{pmatrix} 0 & 0 \\ 0 & I - \tilde{A}_2 \end{pmatrix},$$

where $A_2$ is defined by (2.6). We observe that there cannot be two vectors $w_1$ and $w_2$ of norm one that remain orthogonal as $\epsilon \to 0$ and satisfy $w_i^{\mathrm{T}} X = o(\epsilon)$. For otherwise, there would be a linear combination $w_3 = (0\ \bar{w}_3^{\mathrm{T}})^{\mathrm{T}}$ of $w_1$ and $w_2$ that satisfies $\|w_3\| = 1$ and $w_3^{\mathrm{T}} X = o(\epsilon)$. It then follows that $\bar{w}_3^{\mathrm{T}}(I - \tilde{A}_2) = o(\epsilon)$, contradicting Regularity Condition 3. Since $I - B_y$ is similar to $X$ and the similarity transformation is uniformly bounded, $I - B_y$ likewise cannot have two approximate null vectors that map into quantities of $o(\epsilon)$. We shall establish (3.7) by showing that if $1 - \beta_1 = o(\epsilon)$ then the matrix $I - B_y$ indeed has two such approximate null vectors.

The matrix $I - B_y$ has the form

$$I - B_y = \begin{pmatrix} 1 - \beta_1 & -h^{\mathrm{T}} \\ -g & I - B_2 \end{pmatrix}.$$

Since $(I - B_y)u = 0$, where $u$ is the vector consisting of all ones, if $1 - \beta_1 = o(\epsilon)$ then we must also have $\|h\| = o(\epsilon)$. Thus $w_1 = (1\ 0)^{\mathrm{T}}$ is one $o(\epsilon)$ approximate null vector.

We claim that there is a vector $\bar{w}_2$ with $\|\bar{w}_2\| = 1$ such that $\bar{w}_2^{\mathrm{T}}(I - B_2) = o(\epsilon)$. For otherwise, the perturbation theory in [8, p. 295] would apply to show that there is a left eigenvector of $I - B_y$ of the form $(1\ o(1)$

corresponding to an eigenvalue that is $o(\epsilon)$ in size. Since, by Regularity Condition 3, there is only one such eigenvalue, namely the zero eigenvalue, the eigenvector must be a multiple of $v_y$. But by Regularity Condition 1 the components of $v_y$ cannot approach zero — a contradiction.

Since $(-g\ I - B_2)u = 0$ it follows that $\bar{w}_2^{\mathrm{T}} g = o(\epsilon)$. Thus $w_2 = (0\ \bar{w}_2^{\mathrm{T}})^{\mathrm{T}}$ is a second $o(\epsilon)$ approximate null vector of $I - B_y$. Since $w_1$ and $w_2$ are orthogonal, we have arrived at a contradiction, and equation (3.7) is established.

## REFERENCES


[1] P.-J. Courtois. *Decomposability.* Academic Press, New York, 1977.

[2] J. J. Dongarra, J. R. Bunch, C. B. Moler, and G. W. Stewart. *LINPACK User's Guide.* SIAM, Philadelphia, 1979.

[3] S. Karlin and H. M. Taylor. *A First Course in Stochastic Processes.* Academic Press, New York, second edition, 1975.

[4] D. F. McAllister, G. W. Stewart, and W. J. Stewart. On a Rayleigh–Ritz refinement technique for nearly uncoupled stochastic matrices. *Linear Algebra and Its Applications*, 60:1–25, 1984.

[5] H. A. Simon and A. Ando. Aggregation of variables in dynamic systems. *Econometrica*, 29:111–138, 1961.

[6] B. T. Smith, J. M. Boyle, J. J. Dongarra, B. S. Garbow, Y. Ikebe, V. C. Klema, and C. B. Moler. *Matrix Eigensystem Routines - EISPACK Guide Lecture Notes in Computer Science.* Springer-Verlag, New York, 2nd edition, 1976.

[7] G. W. Stewart. Computable error bounds for aggregated Markov chains. *Journal of the ACM*, 30:271–285, 1983.

[8] G. W. Stewart. *Introduction to Matrix Computations.* Academic Press, New York, 1973.

[9] X. Sun. *A Unified Theory of Nearly Uncoupled Markov Chains.* Ph.D. Thesis, Department of Computer Science, University of Maryland, College Park, MD, 1991.

[10] Y. Takahashi. *A Lumping Method for Numerical Calculations of Stationary Distributions of Markov Chains.* Research Report B-18, Department of Information Sciences, Tokyo Institute of Technology, 1975.

[11] H. Vantliborgh. *The Error of Aggregation in Decomposable Systems.* Report R 453, Philips Research Laboratory, Brussels, Belgium, 1981.

[12] R. S. Varga. *Matrix Iterative Analysis.* Prentice-Hall, Englewood Cliffs, New Jersey, 1962.

# MINIMUM RESIDUAL MODIFICATIONS TO BI-CG AND TO THE PRECONDITIONER

HENK A. VAN DER VORST*

**1. Introduction.** We will be concerned with the iterative solution of the linear system $Ax = b$, where $A$ is a nonsymmetric nonsingular matrix. If a suitable preconditioner is available then the linear system is to be understood as having been preconditioned already.

In Krylov subspace methods one generates a basis for the Krylov subspace $K^i(A; r_0) \equiv \text{span}\{r_0, Ar_0, ..., A^{i-1}r_0\}$, where $r_0 = b - Ax_0$, projects the given system onto this space and solves the (much) smaller projected system. The computational costs for the construction of a full orthogonal basis for $K^i(A; r_0)$ , as is done in GMRES [13], becomes increasingly expensive: it grows like $i^2$. Moreover, all the basis vectors have to be kept in memory, so that the required memory space grows linearly with $i$. The advantage of GMRES is that one obtains the $x_i \in K^i(A; r_0)$, for which $||b - Ax_i||_2$ is minimal.

In order to limit the memory space requirements one usually restarts methods like GMRES after a fixed number of steps ($m$, say). However, with restarting, the convergence behaviour depends strongly on the choice of $m$ for each given problem, and the best strategy seems to be to take $m$ as large as storage space permits [9] (which is still no guarantee for fast convergence for general situations).

One may try to improve the GMRES behaviour by selecting other search directions. In the GMRESR approach, suggested in [18], a variable preconditioner is included. This preconditioning process may consist of a suitable (inner-) iteration method. We will discuss this approach in Section 3.

An alternative for the full orthogonalization approach is to generate a suitable set of basis vectors by short recurrence relations. In Bi-CG [4] one generates two bi-orthogonal sets of basis vectors $\{v_j\}$ and $\{\hat{v}_j\}$, for $K^i(A; r_0)$ and $K^i(A^T; \hat{r}_0)$, respectively, by 3-term recurrence relations similar to those used in CG. We will discuss the choice of $\hat{r}_0$ in Section 2.

But Bi-CG got a dubious reputation because of the possibilities of break-down of the method (which are rare events in practice though), and because there is no minimization property for the residual, which is reflected by a sometimes rather irregular behaviour of the error or the residual. Nevertheless, for many problems the method has appeared to be an efficient solver for relevant problems and at present we see many attempts to improve the method or to repair its deficiencies. Of these attempts we

* Math. Institute, State University of Utrecht, P.O. Box 80.010, NL-3508 TA Utrecht, The Netherlands.

mention the look-ahead strategies to avoid break-down [5], and strategies to obtain a more smooth or even faster convergence behaviour (e.g., QMR in [6]).

Quite another approach has been suggested by Sonneveld [14]. Based upon the observations that $r_i = p_i(A)r_0$ and $\hat{r}_i = p_i(A^T)\hat{r}_0$, where $p_i$ is an $i$-th degree polynomial with constant coefficient 1, and that vectors like $\hat{r}_i$ are only used in innerproducts for the computation of the iteration coefficients, Sonneveld has shown that with the same amount of arithmetical work one can compute approximate solutions $\bar{x}_i$ for which $b - A\bar{x}_i = p_i^2(A)r_0$. In this approach one can avoid to work with $A^T$. The resulting method is now widely known as CGS.

Often one observes for CGS a significantly faster convergence (up to about a factor of two) in terms of CPU-time, than for Bi-CG. This can be explained heuristically by the observation that for $\bar{r}_i$ the operator $p_i(A)$ has been applied twice on $r_0$. On the other hand, since the operator $p_i(A)$ depends strongly on $r_0$, there is no guarantee at all that $p_i(A)$ should have about the same reduction properties for the vector $p_i(A)r_0$ as for $r_0$.

The squaring of the polynomial $p_i$ may result in a quite irregular convergence behaviour of the $\|\bar{r}_i\|_2$'s. In [15] it is explained that the ratio $\|\bar{r}_i\|_2/\|\bar{r}_{i-1}\|_2$ can be very large, even in situations where for the same iteration number this ratio for Bi-CG is smaller than 1. In [16] it is shown by example that very large values of $\|\bar{r}_i\|_2$, with respect to $\|r_0\|_2$, may lead to large errors in the approximated solution $\bar{x}_i$, in the sense that the vector $\bar{r}_i$, obtained through updating in the CGS-process, can be largely different from the true residual $b - A\bar{x}_i$. It is shown that this defect in the method cannot always be repaired by computing the true residual in CGS (which is even cheaper than computing the updated one), since this practice may even prevent convergence of CGS.

Following the ideas of Sonneveld it was shown in [16] that one can also form, at modest costs, approximations for which the residuals can be written as the product of $p_i(A)$ with some other $i$-th degree polynomial, applied to $r_0$. A rather successful result of this approach, the Bi-CGSTAB method, will be discussed in section 2.

**2. The Bi-CGSTAB Method.** In an attempt to damp the irregular convergence behaviour (especially the peaks) of CGS, it was suggested in [16] to construct approximations $\tilde{x}_i$, for which the corresponding residuals can be written as

$$\tilde{r}_i = q_i(A)p_i(A)r_0,$$

where $q_i(t) = (1 - \omega_i t)(1 - \omega_{i-1}t) \cdots (1 - \omega_1 t)$.

Given the values of $\omega_{i-1}, \ldots, \omega_1$, the value of $\omega_i$ is computed as to minimize $\|\tilde{r}_i\|_2$, seen as a function of $\omega_i$. The factor $I - \omega_i A$ can be seen as to represent a minimum residual step (or GMRES(1)) on top of the Bi-CG process and the previously carried out GMRES(1) steps. See also [7] for the interpretation of this process as a product of Bi-CG and repeated GMRES(1), and the further exploitation of this viewpoint.

The resulting method has been given the name Bi-CGSTAB and from the iteration scheme below it can easily be concluded that it requires about the same arithmetical effort per iteration step as CGS (or Bi-CG). With CGS it shares the advantage that there are no operations with $A^T$.

The Bi-CGSTAB algorithm can be represented by the following scheme (which has been taken from [16]):

$x_0$ is an initial guess; $r_0 = b - Ax_0$;
$\hat{r}_0$ is an arbitrary vector, such that
    $(\hat{r}_0, r_0) \neq 0$, e.g., $\hat{r}_0 = r_0$;
$\rho_0 = \alpha = \omega_0 = 1$;
$v_0 = p_0 = 0$;
for $i = 1, 2, 3, \ldots$
    $\rho_i = (\hat{r}_0, r_{i-1}); \beta = (\rho_i/\rho_{i-1})(\alpha/\omega_{i-1})$;
    $p_i = r_{i-1} + \beta(p_{i-1} - \omega_{i-1} v_{i-1})$;
    $v_i = Ap_i$;
    $\alpha = \rho_i/(\hat{r}_0, v_i)$;
    $s = r_{i-1} - \alpha v_i$;   (1)
    $t = As$;
    $\omega_i = (t, s)/(t, t)$;   (2)
    $x_i = x_{i-1} + \alpha p_i + \omega_i s$;
    if $x_i$ is accurate enough then quit;
    $r_i = s - \omega_i t$;
end

(1) The vector $s$ represents the residual after the Bi-CG part of the algorithm. If the norm of this vector is sufficiently small, then the algorithm could be terminated after updating the current approximation for $x$ appropriately: $x_i = x_{i-1} + \alpha p_i$.
(2) The interpretation of Bi-CGSTAB as the product of Bi-CG and GMRES(1) processes reveals two possible causes for breakdown: the underlying Bi-CG algorithm may suffer from breakdown, or the GMRES(1) part may lead to no (or marginal) improvement: $\omega_i = 0$ (or very small). In the latter case the method is unable to construct the recursion formula for the next Bi-CG step (see the formula for $\beta$). In principle one could continue the algorithm with any arbitrary not too small value for $\omega_i$, and admit locally a small growth in the norm of the residual. On the other hand, one might argue that if $\omega_i = 0$, then this indicates situations in which applying Bi-CG is also a risky bussiness. Anyway, in situations where Bi-CG has a chance for breakdown, the risk to encounter such an event with Bi-CGSTAB is certainly not less.

In numerical experiments we have observed a remarkable behaviour of

Bi-CGSTAB in the presence of rounding errors. This will be subject to some discussion in the next subsection.

**2.1. The parameter $\rho$.** The schemes for CGS and Bi-CGSTAB are essentially based upon the reconstruction of the BI-CG iteration coefficients $\alpha$ and $\beta$ from certain innerproducts (these parameters define the polynomial $p_i$). It is obvious that the precision with which these parameters can be determined depends quite critically upon the precision in the parameter $\rho_j$. This value is computed with the vector $r_j$, which has been updated in the previous iteration step as $r_j = s - \omega_j t$. In that expression $s$ represents the residual after the BI-CG part of the iteration procedure and $\omega_j t$ is the minimum residual correction in the $j$-th iteration step. It is clear that when the minimum residual correction $\omega_j t$ is very successful, i.e., when $||r_j|| \ll s$, then $r_j$ may be expected to be less significant than either $s$ or $t$. This suggests to compute $\rho_j$ more accurately as

$$\rho_j = (\bar{r}, r_j) = (\bar{r}, s - \omega_j t) = -\omega_j(\bar{r}, t).$$

This has been suggested in [16] and it is stated there that the improved formula might lead to a reduction in iteration steps. However, this is not in agreement with what we have observed in more realistic situations, in particular we have seen degrading effects in semiconductor modelling problems (see also [1]).

As might be expected, the values of the $\alpha$'s and $\beta$'s, that one would have obtained by carrying out the BI-CG scheme, are now better recovered by the changed Bi-CGSTAB scheme in comparison with the standard scheme. But, surprisingly this does not necessarily lead to an improved convergence behaviour. Even on the contrary, it sometimes helps to compute $\rho_j$ as $\rho_j = (\bar{r}, r_j)$ in reduced precision (e.g., in single precision, while all other computation is done in double precision). An example of this can be seen in [1]: Figure 3.

We have seen also many problems where the difference with the standard algorithm was minimal, but in the problems where a different convergence behaviour occurred, Bi-CGSTAB with the "improved" $\rho_j$ computation usually required more iterations. At present we have the impression that the changes in the convergence behaviour are correlated with a loss of bi-orthogonality in the underlying Bi-CG scheme, but we do not know how that could explain the observed differences.

**2.2. On the choice of the starting vector.** Not so much is known about the convergence of these Bi-CG variants, except for special cases (see, e.g., [5] for error bounds). In [11] the Bi-CG method is derived from dertermining turning points of $f = (\hat{r}_i, A^{-1}r_i)$, with $\hat{r}_i = p_i(A^T)\hat{r}_0$, $r_i = p_i(A)r_0$, and $p_i(0) = 1$. Of course, the behaviour of the functional $f$ is affected by the choice of $\hat{r}_0$. Since the turning points do not, in general, necessarily represent minima, we need not be surprised to see any kind of irregular behaviour in the convergence of Bi-CG.

The most common choice is $\hat{r}_0 = r_0$. If $A$ is symmetric positive definite this leads to the Conjugate Gradients method, by which $(r_i, A^{-1}r_i)$ is minimized over $K^i(A; r_0)$. When $A$ is only symmetric, then, of course, CG may fail, but an alternative would be to minimize $(r_i, r_i)$. This is easily accomplished in the Bi-CG method by selecting $\hat{r}_0 = A^T r_0$, which for $A = A^T$ leads to the conjugate residual method (or minimum residual method). Similarly as with CG, this method can be implemented with only one matrix vector product per iteration step. Note that in the nonsymmetric case it does not make sense to rename Bi-CG for this choice of starting vectors as the Bi-Conjugate Residual method, since formally $\hat{r}_0$ is a free parameter in Bi-CG.

It has been observed in our experiments, for systems of which the matrix $A$ has a few negative eigenvalues, and of which the eigenvalues in the positive right half plane have relatively small imaginary components, that the choice $\hat{r}_0 = A^T r_0$ for Bi-CG leads to considerably smoother convergence behaviour. Also for CGS and Bi-CGSTAB we have seen a smoother, and even faster, convergence behaviour in our experiments.

These observations show that we may expect entirely different convergence behaviours. If $A$ is symmetric indefinite, then with $\hat{r}_0 = A^T r_0$ we are minimizing $(r_i, r_i)$, whereas with $\hat{r}_0 = r_0$ the BI-CG process may even break-down.

There are other ways to play with the choice of $\hat{r}_0$. In [10], [1] it is reported that the choice of a random vector $\hat{r}_0$ may give better convergence for CGS than the choice $\hat{r}_0$. This can be understood heuristically, if we restrict ourselves to symmetric positive systems [1]. If $r_0$ has small components $\gamma_j$ in some eigenvector directions $v_j$ ($Av_j = \lambda_j v_j$, $\|v_j\|_2 = 1$), then we allow $p_i(\lambda_j)$ to take large values for the corresponding eigenvalues. Then it may well be, and this is easily verified for artificial examples, that components like $p_i^2(\lambda_j)\gamma_j$, at some iteration steps, lead to very large values of the norm of this residual, even in situations in which the Bi-CG residual is still decreasing in norm.

With the choice of a random vector for $\hat{r}_0$, the polynomial $p_i$ is constructed with respect to the weights $\hat{\gamma}_j\gamma_j$, and since these weights may be expected to be less unevenly distributed than the $\gamma_j^2$, it is now less likely that $p_i$ will take very large values in some of the eigenvalues.

In [1] this is further exploited in the design of starting strategies for CGS and Bi-CGSTAB for the solution of linear systems arising in Newton's method for nonlinear semiconductor device problems.

**3. Rank-one Updates to the Preconditioner.** A splitting of the matrix $A = K - R$ leads immediately to the basic iteration $Kx_i = b + Rx_{i-1}$. In [2] it is suggested to improve the convergence behaviour of the basic iteration by adapting the matrix splitting, using information obtained in the iteration process. We will give the flavour of this method since it turns out that, when combined with a minimum residual approach, it has an interesting relation with GMRES-type methods (like GMRES, ORTHODIR and GENCG). This approach is followed in [17] for the construction of new classes of GMRES-like methods, that can be used as cheap alternatives for the increasingly expensive full GMRES method.

Assume that the matrix splitting in the $k$-th iteration step is given by $A = H_k^{-1} - R_k$, then we obtain the iteration formula

$$x_k = x_{k-1} + H_k r_{k-1} \quad \text{with} \quad r_k = b - Ax_k.$$

The idea is to construct $H_k$ by a suitable rank-one update to $H_{k-1}$:

$$H_k = H_{k-1} + u_{k-1} v_{k-1}^T,$$

which leads to

$$x_k = x_{k-1} + (H_{k-1} + u_{k-1} v_{k-1}^T) r_{k-1} \tag{3.1}$$

or

$$\begin{aligned} r_k &= r_{k-1} - A(H_{k-1} + u_{k-1} v_{k-1}^T) r_{k-1} \\ &= (I - AH_{k-1}) r_{k-1} - \mu_{k-1} A u_{k-1}, \end{aligned} \tag{3.2}$$

with $\mu_{k-1} = v_{k-1}^T r_{k-1}$. Given a suitable choice for $u_{k-1}$, the constant $\mu_{k-1}$ is taken such that it minimizes the 2-norm of $r_k$:

$$\mu_{k-1} = \frac{1}{||A\bar{u}_{k-1}||_2^2} (A\bar{u}_{k-1})^T (I - AH_{k-1}) r_{k-1}.$$

Since $v_{k-1}$ satisfies the relation $\mu_{k-1} = v_{k-1}^T r_{k-1}$, we have the following obvious choice:

$$v_{k-1} = \frac{1}{||Au_{k-1}||_2^2} (I - AH_{k-1})^T A u_{k-1} \tag{3.3}$$

(note that from the minimization property we have that $r_k \perp Au_{k-1}$).

In principle the implementation of the method is quite straight forward, but note that the computation of $r_{k-1}$ and $v_{k-1}$ could be quite expensive. This would make the method unpractical. Also the updated splitting is most likely a dense matrix if we carry out the updates explicitly. We will show, still following the lines set forth in [2], that the method can be implemented much more efficiently.

We define

1. $c_k = \frac{1}{||Au_k||_2} Au_k$
2. $E_k = I - AH_k$

From (3.2) we have that $r_k = E_k r_{k-1}$. Furthermore:

$$\begin{aligned}(3.4)\quad E_k &= I - AH_k = I - AH_{k-1} - Au_{k-1}v_{k-1}^T \\ &= I - AH_{k-1} - Au_{k-1}(Au_{k-1})^T(I - AH_{k-1})\frac{1}{\|Au_k\|_2^2} \\ &= (I - c_{k-1}c_{k-1}^T)E_{k-1} \\ &= \prod_{i=0}^{k-1}(I - c_i c_i^T)E_0.\end{aligned}$$

We see that the operator $E_k$ has the following effect on a vector. The vector is multiplied by $E_0$ and then successively orthogonalized with respect to $c_0$, ..., $c_{k-1}$. If the set $\{c_j\}$ is not orthogonal, then the result is not necessarily orthogonal to all these vectors. The choice for $u_k$, suggested in [2], leads to an orthonormal set $\{c_j\}$. For general choices, one has to make the $c_j$'s explicitly orthonormal in order to ensure convergence.

The actual implementation is based on the above properties. Given $r_k$ we compute $r_{k+1}$ as follows (and we update $x_k$ in the corresponding way):

$$r_{k+1} = E_{k+1}r_k.$$

With $\xi^{(0)} = E_0 r_k$ we first compute (with the $c_j$ from previous steps):

$$E_k r_k = \xi^{(k)} \equiv \prod_{j=0}^{k-1}(I - c_j c_j^T)\xi^{(0)}.$$

The computed updates $-c_j^T\xi^{(0)}c_j$ for $r_{k+1}$ correspond to updates

$$c_j^T\xi^{(0)}A^{-1}c_j = c_j^T\xi^{(0)}u_j/\|Au_j\|_2$$

for $x_{j+1}$. These updates are in the scheme, given below, represented by $\eta^{(k)}$.

Having selected a suitable $u_k$ (we will discuss this choice later), we have to make $Au_k \sim c_k$ orthogonal with respect to $c_0$, ..., $c_{k-1}$, and to update $u_k$ accordingly. Once we have done that we can do the final update step with $H_{k+1}$, and we can update both $x_k$ and $r_k$ by the corrections resulting from the new direction $c_k$.

The resulting scheme for the $k$-th iteration step becomes:

1. $\xi^{(0)} = (I - AH_0)r_k$; $\eta^{(0)} = H_0 r_k$;
   for $i = 0, ..., k-1$ do
   $\alpha_i = c_i^T\xi^{(i)}$; $\xi^{(i+1)} = \xi^{(i)} - \alpha_i c_i$; $\eta^{(i+1)} = \eta^{(i)} + \alpha_i u_i$;
2. Select $u_k^{(0)} \equiv u_k$; $c_k^{(0)} = Au_k^{(0)}$;
   for $i = 0, ..., k-1$ do
   $\beta_i = -c_i^T c_k^{(i)}$; $c_k^{(i+1)} = c_k^{(i)} + \beta_i c_i$; $u_k^{(i+1)} = u_k^{(i)} + \beta_i u_i$;

$c_k = c_k^{(k)}/\|c_k^{(k)}\|_2$; $u_k = u_k^{(k)}/\|c_k^{(k)}\|_2$;
3. $x_{k+1} = x_k + \eta^{(k)} + u_k c_k^T \xi^{(k)}$;
$r_{k+1} = (I - c_k c_k^T)\xi^{(k)}$;

Note that in the above implementation we have 'only' two matrix vector products per iteration step.

In [2] it is suggested to take $u_k = H_k E_k r_k$. This is an approximation for the vector $u_k = A^{-1}E_k r_k$, with which the iteration would terminate after one step. The resulting method has been analysed in [19]. It requires about twice the amount of arithmetic in GMRES per iteration step. Since in many cases only half the number of iterations are required as with GMRES, the method is potentially more efficient (the work in GMRES grows quadratically with the iteration count).

With the choices $H_0 = 0$, and $u_k = H_k r_k$, we obtain GENCG [3] (for a proof of this see [19]), which is algebraically equivalent to GMRES and ORTHODIR [13]. In this case we have that $E_k r_k = r_k$, and hence we we may skip part 1 of the above algorithm, and set $\xi^{(k)} = r_k$, $\eta^{(k)} = 0$, $u_k = \xi^{(k)}$. The result can be seen as a different formulation of GMRES in which we can obtain explicit formulas for the updated preconditioner (i.e., the inverse of $A$ is approximated increasingly well): the update for $H_k$ is $u_k c_k^T E_k$ and the sum of these updates gives an approximation for $A^{-1}$.

In this GMRES-variant we are still free to take other $u_k$. Note that the factor $H_k$ in the choice for $u_k$ was introduced as an approximation for the actually desired $A^{-1}$ With $u_k = A^{-1}r_k$, we would have that $r_k = E_{k-1}r_{k-1} - \mu_{k-1}r_{k-1} = 0$ for the minimizing $\mu_{k-1}$. This viewpoint suggests other approximations for the inverse (with respect to the given residual $r_{k-1}$), e.g., to apply a few steps of GMRES to $Ay = r_{k-1}$, and to take the resulting approximation for $y$ as $u_k$. This leads to the so-called GMRESR family of nested methods (for details see [17]).

The success of the above method depends, of course, on the choice of $u_k$. Problems arise if the new $c_k$ is linearly dependent of the previous vectors, since in that case we are not expanding our search space and the method may break down. This situation may happen if we solve $Ay = r_{k-1}$, starting with $y_0 = 0$, and when the selected number of GMRES iteration steps , $m$ say, does not lead to any improvement, i.e. $y_m = 0$. In [17] it is shown that the method is robust if we carry out at least one iteration step with LSQR [12] for the approximation of $y$.

In our experiments we have seen that it may be wise to include more often LSQR steps in the process, not only in stagnating inner iteration processes, but also in situations where the inner iteration converges only very slowly.

## REFERENCES

[1] M. Driessen and H. van der Vorst. Bi-CGSTAB in semiconductor modelling. In Wolfgang Fichtner and Dölf Aemmer, editors, *Simulation of Semiconductor Devices and Processes, Vol 4*, pages 45–54, Konstanz, 1991. Hartung-Gorre Verlag.

[2] T. Eirola and O. Nevanlinna. Accelerating with rank-one updates. *Lin. Alg. and its Appl.*, 121:511–520, 1989.

[3] S. C. Eisenstat, H. C. Elman, and M. H. Schultz. Variational iterative methods for nonsymmetric systems of linear equations. *SIAM J. Numer. Anal.*, 20:345–357, 1983.

[4] R. Fletcher. *Conjugate gradient methods for indefinite systems*, volume 506 of *Lecture Notes Math.*, pages 73–89. Springer-Verlag, Berlin, 1976.

[5] R. W. Freund and N. M. Nachtigal. An implementation of the look-ahead Lanczos algorithm for non-Hermitian matrices, part 2. Technical Report 90.46, RIACS, NASA Ames Research Center, 1990.

[6] R. W. Freund and N. M. Nachtigal. QMR: a quasi-minimal residual method for non-Hermitian linear systems. Technical Report 91-05, IPS ETH, Zürich, 1991.

[7] M. H. Gutknecht. Variants of BICGSTAB for matrices with complex spectrum. Technical Report 91-14, IPS ETH, Zürich, 1991.

[8] M. H. Gutknecht. A completed theory of the unsymmetric Lanczos process and related algorithms, Part I. *SIAM J. Matrix Anal. Appl.*, 1992.

[9] Y. Huang and H. A. van der Vorst. Some observations on the convergence behaviour of GMRES. Technical Report 89-09, Delft University of Technology, Faculty of Tech. Math., 1989.

[10] W. Youbert. Iterative Methods for the Solution of Nonsymmetric Systems of Linear Equations. Report CNA-242, Center for Numerical Analysis, University of Texas at Austin, 1990.

[11] G. Markham. Conjugate gradient type methods for indefinite, asymmetric, and complex systems. *IMA J. of Num. An.*, 10:155–170, 1990.

[12] C. C. Paige and M. A. Saunders. LSQR: An algorithm for sparse linear equations and sparse least squares. *ACM Trans. Math. Soft.*, 8:43–71, 1982.

[13] Y. Saad and M. H. Schultz. GMRES: a generalized minimal residual algorithm for solving nonsymmetric linear systems. *SIAM J. Sci. Statist. Comput.*, 7:856–869, 1986.

[14] P. Sonneveld. CGS: a fast Lanczos-type solver for nonsymmetric linear systems. *SIAM J. Sci. Statist. Comput.*, 10:36–52, 1989.

[15] H. A. van der Vorst. The convergence behaviour of preconditioned CG and CG-S in the presence of rounding errors. In O. Axelsson and L. Yu. Kolotilina, editors, *Preconditioned Conjugate Gradient Methods*, Berlin, 1990. Nijmegen 1989, Springer Verlag. Lecture Notes in Mathematics 1457.

[16] H. A. van der Vorst. Bi-CGSTAB: A fast and smoothly converging variant of Bi-CG for the solution of non-symmetric linear systems. *SIAM J. Sci. Statist. Comput.*, 13:631–644, 1992.

[17] H. A. van der Vorst and C. Vuik. GMRESR: A family of nested GMRES methods. Technical Report 91-80, Delft University of Technology, Faculty of Tech. Math., 1991.

[18] H. A. van der Vorst and C. Vuik. The rate of convergence of the GMRES method. Technical Report 654, University of Utrecht, Dept. of Math., 1991.

[19] C. Vuik and H.A. van der Vorst. A comparison of some GMRES-like methods. *Lin. Alg. and its Appl.*, 160:131–162, 1992.